Hazardous Waste Management

All companion files for this title are available by visiting the website for the book at *sciendo.com/book/9781501524264* and by clicking on the "COMPANION FILES" tab.

Hazardous Waste Management:

An Introduction

Third Edition

Cliff VanGuilder

Mercury Learning and Information
Boston, Massachusetts

MERCURY LEARNING AND INFORMATION
121 High Street, 3rd Floor
Boston, MA 02110
info@merclearning.com

C. VanGuilder. *Hazardous Waste Management: An Introduction, Third Edition.*
ISBN: 978-1-5015-2426-4

Library of Congress Control Number: 2025940813
242526321 This book is printed on acid-free paper in the United States of America.

All companion files for this title are available by visiting the website for the book at *sciendo.com/book/9781501524264* and by clicking on the "COMPANION FILES" tab.

CONTENTS

ACKNOWLEDGMENTS

To my Family, thank you for all your love, patience, and support.

To my church family, thanks for your prayers and moral support.

To Tess Healey, thanks for your friendship, great technical review, and sound advice.

To Howard Brezner and the rest of my former team, thanks for the great experiences.

To Lauren Sherman, thanks for your friendship and technical review.

To my excellent team at VanGuilder Engineering and ARCH Environmental, thanks for the great work you do that inspires me every day.

To Bruce Bergwall, thanks for putting me in touch with the right people.

To David Pallai, thanks for all your help. Enjoy your well-earned retirement.

INTRODUCTION

The first edition of this book, *Hazardous Waste Management: An Introduction* was published in 2012, and several portions will be out of date.

The second edition, published in 2018, was very timely, because in 2014 and 2015, the United States Environmental Protection Agency (USEPA) introduced three major revisions to the Solid and Hazardous Waste Regulations, those being the first major revisions since 1990. These revised standards are taking effect across all states (on schedules determined by each state) authorized by the USEPA to administer their own programs.

These changes were implemented by the USEPA for a variety of reasons, with the majority of the solid waste regulations intended to tighten up the regulation of recycling facilities, including hazardous waste recycling. The bulk of the hazardous waste regulation changes were to simplify the regulation of the smallest facilities that generate hazardous waste and to recognize and help businesses that generate larger quantities of hazardous waste on a one-time or infrequent (episodic) basis.

The second edition also included a description and discussion of a separate regulatory change in the hazardous waste arena, promulgated by the USEPA on December 1, 2008. This action added a new Subpart K, which applies to colleges and universities as well as teaching hospitals and nonprofit research institutes that are either owned by or formally affiliated with a college or university. It was not included in the 2012 version of this book because it had not yet been implemented in the individual states.

One very significant exception to these state-determined effective rule dates is the e-Manifest Rule, a separate rule that took effect June 30, 2018. USEPA had indicated a desire for a national tracking system for all hazardous wastes since 2011, because all shipments are tracked by each state and are therefore difficult to track nationally. A new law, called the Hazardous Waste Electronic Manifest Establishment Act, was signed into law October 5, 2012. The purpose of this law was to establish a national tracking system for all hazardous waste systems, and for the costs of the development and operation system to be recovered from fees charged to the users of the manifests. The implementation of this system took the USEPA seven years.

A more detailed description of the e-Manifest system and the user fees can be found at *https://www.epa.gov/e-manifest/learn-about-hazardous-waste-electronic-manifest-system-e-manifest* and *https://www.epa.gov/e-manifest/final-rule-user-fees-electronic-hazardous-waste-manifest-system-e-manifest-and-amendments* and are discussed in more detail in Chapter 3.

This third edition was written to introduce recent updates to the hazardous waste program, including proposed changes to the Resource Conservation and Recovery Act regulations (RCRA).

One proposed change is to add nine per- and polyfluoroalkyl substances (PFAS), their salts, and their structural isomers to the list of hazardous constituents in Title 40 of the Code of Federal Regulations Part 261 Appendix VIII. These nine PFAS are:

1. Perfluorooctanoic acid
2. Perfluorooctanesulfonic acid
3. Perfluorobutanesulfonic acid
4. Hexafluoropropylene oxide-dimer acid
5. Perfluorononanoic acid
6. Perfluorohexanesulfonic acid
7. Perfluorodecanoic acid
8. Perfluorohexanoic acid
9. Perfluorobutanoic acid

These chemicals are a cause of concern due to their widespread use in personal care products, including cosmetics, food packaging, and waterproofing of clothing and their links to a variety of health concerns. Per the EPA, "current scientific research suggests that exposure to certain PFAS may lead to adverse health outcomes."

A second related proposed rule would broaden the definition of "hazardous waste" applicable to RCRA corrective action. The proposed rule would implement EPA's interpretation that RCRA provides it authority to address and take corrective action for releases of any substances that meet the broad statutory definition of "hazardous waste." Under RCRA's Corrective Action Program, EPA and authorized states manage investigations and cleanups of hazardous waste and hazardous constituents at treatment, storage, and disposal facilities (TSDFs). Hazardous waste sites are often brought into RCRA's Corrective Action Program after there is a confirmed release of a hazardous waste or hazardous constituent, or if the regulatory agency is considering a TSDF permit application.

A third rule covered in this edition is a change to the Federal Resource Conservation and Recovery Act (RCRA) regulations effective February 7, 2020. The EPA added hazardous waste aerosol cans to the universal waste program. This change provided a clear, protective system for managing discarded aerosol cans. According to the USEPA, "the streamlined universal waste regulations are expected to ease regulatory burdens on retail stores and others that discard hazardous waste aerosol cans; promote the collection and recycling of these cans; and encourage the development of municipal and commercial programs to reduce the quantity of these wastes going to municipal solid waste landfills or combustors" [EPA 19].

This book starts with a history of the solid waste and hazardous waste regulations in the United States, including some of my personal experiences in hazardous waste cleanups and development of some of the regulations. Although we have learned valuable lessons from past tragedies like Love Canal, the more recent catastrophic earthquake and tsunami, followed by radioactive releases from the Daiichi nuclear power plant reactors in Fukushima, Japan, show us that we need to continually improve our preparedness for hazardous waste and hazardous materials emergencies worldwide. The recent revelation concerning the adverse health effects of PFAS has had a profound effect on the universe of environmental regulation. When these compounds were discovered and put into the mass production of household products like Teflon cookware, waterproof clothing, food wraps, and countless other products, they were seen as incredibly beneficial. No one thought they would present the myriad challenges and issues they do today.

This text was written for four main purposes: the first being to introduce the reader to the technical field of hazardous waste management; the second to help the reader understand the myriad federal and state hazardous waste regulations; the third to offer some cost-saving measures for companies who do not completely understand the regulations; and the fourth to show how significant relief can be achieved by completely acceptable measures like neutralization of corrosive wastes under certain circumstances. It was impossible to separate these four subjects because in the United States, the regulations are frequently written with prescribed treatment standards and, in some cases, prescribed treatment technologies for various hazardous waste streams. This book will explain the history of the regulations and how the regulations became so prescriptive for certain wastes.

The hazardous waste regulations in the United States are often confusing, with a large number of references, cross-references, sections, subsections, and sub-subsections. With the assistance of some very patient and talented coworkers in the New York State Department of Environmental Conservation (NYSDEC), colleagues from other states, and friends at the United States Environmental Protection Agency (USEPA), along with untold hours of reading, I was able arrive at a level of understanding that allowed me to feel comfortable writing this book. That process took from 1983–1989, while I was in charge of the development of New York State's hazardous waste regulations and served on several committees to help the USEPA develop the federal hazardous waste regulations.

From 1989 to 2009, I was able to apply these regulations as a supervising hazardous waste inspector, certified by the USEPA and the NYSDEC. I conducted hundreds of hazardous waste inspections across New York State and reviewed over one thousand inspections prepared by my staff. I supervised several dedicated and talented inspectors, reviewing their completed work for accuracy and completeness. I trained several hazardous waste inspectors, prepared them for certification, and conducted a certification exam, deciding whether they should be certified as hazardous waste inspectors.

During the time I was a supervising hazardous waste inspector, my very talented and dedicated staff developed a computerized hazardous waste inspection program that produced a finished copy of inspection reports, eliminating untold hours of paperwork and saving thousands of pages of printed paper. The program my staff developed in our office produced a finished report, and, based on any violations found in the inspection, also produced the correspondence to the facility inspected. This correspondence would range from a letter stating no violations were found, to a list of violations that needed to be resolved within a specific time frame (usually 30 days), to a complaint with proposed penalties.

The inspections were conducted at all categories and sizes of hazardous waste generators in New York State, along with hazardous waste treatment, storage, and disposal facilities and hazardous waste transfer facilities. When I conducted inspections, the owner or environmental manager of the company would frequently ask, "What do I need to do to pass hazardous waste inspections?" I would reply that they had to prove they were in compliance with all of the pertinent hazardous waste laws rules and regulations. This was accomplished in two steps: (a) determine the proper regulatory category(ies) of the facility and (b) make sure the facility complied with all the questions contained in the pertinent hazardous waste compliance checklists, which the inspector would supply at the end of the inspection.

Several textbooks have been written about hazardous waste management. Some of these books deal with cleanup of contamination from past practices. Others attempt to explain the hazardous waste regulations from an environmental perspective. This book is written as a practical guide for everyone, including professors, students, business owners, environmental professionals, and lawyers, so they may understand the hazardous waste regulations from both a technical and regulatory compliance perspective.

The book includes a discussion of the hierarchy of hazardous waste management. This is followed by straightforward instructions on how to determine whether a facility generates or otherwise manages hazardous waste. After these instructions, the book describes several hazardous waste management methods and treatment technologies. The book then provides a list of requirements that must be met to comply with the pertinent regulatory citations. The book gives the regulatory citation from the federal Code of Federal Regulations (CFR), provides an explanation of the terms in the citation, and gives the appropriate compliance question(s) that would be asked by a compliance inspector. If the readers start with a thorough and accurate waste determination and follow the instructions in the book and the companion files carefully, they should be able to develop a basic comprehensive understanding of hazardous waste management in the United States. They should also develop a good understanding of the basic regulatory requirements for hazardous waste management and know where to look for the appropriate U.S. rules and regulations.

Please note this book is about the hazardous waste management rules and regulations in the United States, and that the rules and regulations in each of the states may be stricter than the federal rules and regulations. The rules and regulations for countries may be very different from U.S. standards.

This book contains several case studies from actual inspections that provide questions designed to challenge readers on interpreting the regulations. The identities of the facilities and the employees in the case studies are kept anonymous unless the information was made public by the regulating agency.

For professors, I have provided instructional slides from each chapter in Microsoft PowerPoint. I have also included companion files with questions and answers for chapter quizzes.

All companion files for this title are available by visiting the website for the book at *sciendo.com/book/9781501524264* and by clicking on the "COMPANION FILES" tab.

Owners of businesses and environmental managers face a daunting task. They are expected to be familiar with a myriad of environmental regulations and to comply with them all. The hazardous waste regulatory program is arguably the most complicated environmental regulatory program, in terms of understanding the language and the requirements. This book is written to translate the language and intention of these complex hazardous waste regulatory requirements into terms the business owners and environmental regulators can understand.

The book also offers options for potential regulatory relief and cost savings in areas like neutralization of certain corrosive hazardous wastes, identification of certain hazardous wastes as universal wastes, and other areas.

Since my retirement with the NYSDEC in 2009, I have continued my work in the environmental engineering and compliance arena as the owner of two consulting firms, ARCH Environmental Compliance Service Associates LLC and VanGuilder Engineering PLLC, both of which help clients deal with environmental compliance.

Both of my firms offer compliance training in the form of preliminary compliance audits to prepare for regulatory inspections, written plans for compliance, and environmental and civil/structural engineering design.

ARCH Environmental Compliance Service Associates LLC and VanGuilder Engineering PLLC are both located in upstate New York at 333 Kingsley Rd, Burnt Hills, NY, and any questions about these firms can be made at (518) 491-3287.

Cliff VanGuilder, P.E.
Malta, New York
August 2025

REFERENCE

[EPA 19] U.S. Environmental Protection Agency. *Increasing recycling: Adding aerosol cans to the universal waste regulations* EPA 2019-25674 (84 FR 67202). December 9, 2019, online at https://www.federalregister.gov/documents/2019/12/09/2019-25674/increasing-recycling-adding-aerosol-cans-to-the-universal-waste-regulations.

CHAPTER 1

A Brief History of Hazardous Waste

In This Chapter

- Introduction to and definitions of solid waste and hazardous waste
- The history of solid and hazardous waste management
- The first U.S. solid waste law
- The first U.S. hazardous waste laws
- The fundamental principles of hazardous waste management
- Examination of Fukushima Disaster

1.1 INTRODUCTION TO HAZARDOUS WASTE

This chapter introduces readers to hazardous waste management in the United States or, more precisely, the history of solid and hazardous waste, the identification of solid and hazardous waste, and U.S. hazardous waste policy and regulations. Although technical discussions of solid waste, hazardous waste, wastewater, and air pollution control are included in this text, a substantial portion of this book is dedicated to outlining and explaining the hazardous waste regulations in the United States and helping the reader to understand the steps required to comply with them.

Definitions of Solid Waste and Hazardous Waste

A discussion of the history of solid waste and hazardous waste laws, rules, and regulations will begin with definitions of the terms.

Solid Waste Definition

The United States Congress defined *solid waste* as "any garbage, refuse, sludge from a wastewater treatment plant, water supply treatment plant, or air pollution control facility, and other discarded material, including solid, liquid, semisolid, or contained gaseous material resulting from industrial, commercial, mining, agricultural operations, and community activities" [SWDA 65]. The Solid Waste Disposal Act became law on October 20, 1965.

As discussed in the Foreword and described in more detail later in this chapter, the USEPA promulgated changes to the Solid Waste Regulations on December 10, 2014, primarily to tighten the requirements on recyclers of solid and hazardous waste. These changes have been promulgated in most authorized states currently.

NOTE *This definition is confusing from a scientific perspective; these wastes are not necessarily in a solid phase—they can be solid, liquid, or a contained gas.*

Hazardous Waste Definition

Hazardous waste is a solid waste that poses substantial or potential threats to public health or the environment. Figure 1.1 displays an example of a hazardous waste label.

Hazardous waste
FEDERAL LAW PROHIBITS IMPROPER DISPOSAL
If found, contact the nearest police or public safety authority, and the
Washington State Department of Ecology or the Environmental Protection Agency

Accumulation Start Date:
Reportable Quantities (RQ): lbs
40 CFR Subchapter J. Part 302, Table 302.4
Manifest Document #:
Emergency Response Guide #:
EPA Waste Code(s) and/or Characteristic(s)

Generator Name:
Address:
City:
State:
Zip:
EPA ID #:

EPA/DOT Shipping Name:
Hazard Class:
UN/NA #:
Packing Group (PG):

In the event of a spill or release of this hazardous waste, contact the US Coast Guard
National Response Center at 1-800-424-8802 for information and assistance.

FIGURE 1.1 Hazardous waste label (Washington State Department of Ecology at *https://ecology.wa.gov/Regulations-Permits/Guidance-technical-assistance/Dangerous-waste-guidance/Dangerous-waste-basics/Label-dangerous-waste/Print-free-labels*).

Hazardous wastes fall into two major categories: characteristic wastes and listed wastes.

- *Characteristic hazardous wastes* are solid wastes that are known or tested to exhibit a hazardous trait such as:
  - Ignitability (i.e., flammable)
  - Reactivity (reacts vigorously when exposed to water, heat, or pressure)

   - Corrosivity (strong acids or bases)
   - Toxicity (fails test for toxicity)
- *Listed hazardous wastes* are materials specifically listed by the EPA (Environmental Protection Agency) or state as a hazardous waste. Hazardous wastes listed by EPA fall into two major categories:
   - Process wastes from general activities (F-listed) and from specific industrial processes (K-list)
   - Unused or off-specification chemicals, container residues, and spill cleanup residues of acute hazardous waste chemicals (P-listed) and other chemicals (U-list)

The complex official USEPA (United States Environmental Protection Agency) regulatory definitions of solid waste and hazardous waste are discussed later in this chapter, together with plain-language interpretations.

1.2 HISTORY OF SOLID AND HAZARDOUS WASTE MANAGEMENT

The hazardous waste laws, rules, and regulations were generated in reaction to a series of adverse environmental circumstances that occurred over several years, dating as far back as the early twentieth century. Before we look at the hazardous waste problem, we should first examine the history of solid waste management.

The Waste Management Problem

As long as humans have inhabited the Earth, there have been problems with waste disposal. The first documented garbage problems in a populated area were in Rome, Italy.

Side Note

"Roman rubbish was often left to collect in alleys between buildings in the poorer districts of the city. It sometimes became so thick that stepping stones were needed. 'Unfortunately, its functions did not include house-to-house garbage collection, and this led to indiscriminate refuse dumping, even to the heedless tossing of trash from windows' (Casson, Lionel. *Everyday Life in Ancient Rome*, revised and expanded edition. Baltimore: The Johns Hopkins University Press, 1998. p 40.) As a consequence the street level in the city rose, as new buildings were constructed on top of rubble and rubbish" [WIKI 11a].

As time progressed and people became aware of the hazards, human waste was disposed in locations away from populated areas, usually in low areas, and, where possible, away from populated areas. These "garbage dumps" were health nuisances because they generated odors and attracted vermin that could spread diseases.

Health Hazards

These early dumps generally were not contained by any liners, leachate collection systems, or groundwater barriers. Some of the contaminated leachate from the dumps eventually leaked out and polluted groundwater and surface water, creating health problems, but the source was not always immediately obvious. Most dumps did not receive any kind of regular cover material, such as soil, so numerous birds, insects, rodents, and other vermin would go to the dumps to eat. These vermin became vectors of diseases when they left the dumps and migrated to populated areas. Fires were a problem at many of these dumps, sometimes purposely set by humans to reduce the volume of the waste, at other times to drive off the vermin, and sometimes from spontaneous combustion of materials in the landfills. The odors from these dumps were a nuisance, and the smoke and fumes from fires created air pollution that was annoying and unhealthy for the people living or working downwind. An additional problem occurred when moderate or heavy winds blew waste papers, grocery bags, and other light materials from uncovered dumps into cities, villages, suburbs, towns, or onto water bodies. The image in Figure 1.2 is representative of these dump sites.

FIGURE 1.2 Landfill photo (from town of Colonie, New York).

1.3 FIRST U.S. SOLID WASTE LAW

When the Industrial Revolution began in the eighteenth century, the waste management practices adopted by manufacturing facilities were similar to those of early dump sites; sometimes industrial wastes were dumped in the same places where human wastes were disposed, and sometimes individual industrial waste disposal areas were created. In the meantime, while laws were being written to better manage solid waste, industrial waste dumps were creating problems of greater proportions and potential hazards.

In 1965, in reaction to the public outcry concerning these poor solid-waste management practices in the United States, Congress passed the Solid Waste Disposal Act (SWDA). This law outlined environmentally responsible methods for getting rid of trash at household, municipal, commercial, and industrial levels. Wastes described in the Solid Waste Disposal Act were both hazardous and nonhazardous.

In its original form, the SWDA was an attempt to address the solid waste problems confronting the nation through a series of research projects, investigations, experiments, training, demonstrations, surveys, and studies. Congress indicated two reasons for the necessity of the SWDA:

- Advancements in technology resulted in the creation of vastly increased amounts and types of wastes than there had been in the past.
- Rapid growth in the nation's metropolitan areas had caused these areas to experience significant financial, managerial, and technical problems associated with waste disposal.

Over the next ten years, it became evident that the SWDA was not effective in resolving the solid and hazardous waste disposal issues facing the country.

One very good aspect of the passage of the SWDA was that it marked the beginning of a series of laws and regulations that emphasized clean air and resource management. It focused on researching the pollution and waste management problems during that era and resulted in trained experts that could now work to improve environmental waste management and disposal at the state and local government level.

The Love Canal Tragedy

One of the first industrial dumps to make national news that resulted in further legislation was the Love Canal hazardous waste site in Niagara Falls, New York. The Love Canal was an abandoned water power supply canal built as part of a "model city" that was proposed to be built at the eastern edge of Niagara Falls, a project envisioned by Colonel William Love early in the twentieth century. In the 1920s, after Colonel Love's dream failed to materialize, the canal became a dump site for the city of Niagara Falls, and the city regularly unloaded its municipal refuse into the canal. In the 1940s, the U.S. Army began using the site to dump wastes from the World War II war effort, including wastes from the Manhattan Project.

By the 1940s, Hooker Electrochemical Company (later known as Hooker Chemical Company), founded by Elon Hooker, began searching for a place to dump the chemical waste it was producing. In 1942, Hooker was granted permission by the Niagara Power and Development Company to dump wastes into the canal. The canal was drained and lined with thick clay. Into this site Hooker began dumping industrial waste in various containers. The city of Niagara Falls and the U.S. Army continued the dumping of refuse, along with Hooker, for about six years.

In 1948, after World War II had ended and the city of Niagara Falls had ended self-sufficient disposal of refuse, Hooker became the sole user and owner of the site. This dump site was in operation until 1953 [WIKI 11b].

The dump site must have seemed to be a better-than-average location for a landfill, from an engineering and geological perspective, because it was located in deep clay soils, and its bottom and sides were lined with clay. The dump was also covered with clay soils after it was closed. Humans were not exposed to this dump until the site was purchased by the Niagara Falls School Board in 1953 to build the 99th Street School. The school was built, and eventually the remaining land was developed for housing, which was when the problems started to come to light.

Love Canal was not as large as many other industrial dumps across the nation, or even in New York State. It was, however, the first industrial dump site where houses were built in close proximity to an industrial landfill, and the occupants of these houses were likely exposed to high levels of toxic chemicals. Figure 1.3 shows houses that were built close to the Love Canal dump site.

The Hooker Chemical Company was not entirely to blame for the Love Canal tragedy. As mentioned earlier, the dump site was also used by the city of Niagara Falls and the U.S. Army.

Side Note

The residents who bought houses near the Love Canal dump site in the 1950s and 1960s had no idea that they were going to be living next to an inactive hazardous waste cleanup site, or that they would be potentially exposed to the 30,000+ chemicals that had been dumped there. After the New York State Health Department discovered the extent of the contamination, the State of New York offered to buy the affected houses at a fraction of their assessed values. Some (about 900) homeowners accepted the offers and moved out, but some residents (about 90 homes) decided to stay, some houses within a few feet of the highly contaminated dump. A few of these houses are still occupied today. They were allowed to stay because it was their property and the houses were supplied with public drinking water.

A great deal has been written about the history of the Love Canal dump site. Although many of the books point to Hooker Chemical alone as the villain in this tragedy, there is evidence to suggest Hooker Chemical acted responsibly and a different culprit caused the Love Canal area to be developed into residential housing. A February 1981 article in *Reason* magazine thoroughly chronicles the true Love Canal story [Zuesse 81].

Another informative article was written on the thirtieth anniversary of the discovery of the problems at the Love Canal dumpsite [Engelhaupt 08].

FIGURE 1.3 Love Canal homes (from EPA Website *https://search.yahoo.com/yhs/search?fr2=p%3ads%2cv%3aomn%2cm%3asa%2cbrws%3achrome%2cpos%3a1&hspart=reb&hsimp=yhs-ext_onelaunch&p=epa+love+canal+photos&type=0_1025_102_1082_106_230510*). (Accessed April 2025)

1.4 FIRST U.S. HAZARDOUS WASTE LAWS

The Solid Waste Disposal Act was written in 1965 to deal primarily with solid waste from municipal sources. Although the SWDA mentioned industrial wastes, the problems associated with Love Canal and other industrial waste sites near residences had not yet become evident. The outcry from the affected citizens at these sites made national news in the 1970s, and Congress acted.

The Environmental Protection Agency was formed in 1970.

Ramification of Love Canal: The Resource, Conservation and Recovery Act (RCRA)

As a result of the political fallout of Love Canal and other industrial dump sites, Congress passed the Resource Conservation and Recovery Act in 1976 (RCRA). This law was written to better regulate hazardous waste under RCRA Subtitle C and solid waste under RCRA Subtitle D. The hazardous waste portion of the law (Subtitle C) created a national "cradle to grave" hazardous waste management tracking (manifest) program to deal with the nation's annual production and shipping of hazardous waste. Among many other restrictions and obligations, it also required generators of hazardous waste to file biennial reports.

The solid waste part of the law (Subtitle D) dealt with municipal waste disposal.

Authorization of States

One important provision of the law was the ability for states to obtain authorization to develop and run their own hazardous waste programs, with federal funding support and oversight by the USEPA. To obtain authorization, each state had to pass its own separate law(s) to regulate the management of waste and promulgate regulations that were at least as stringent as the federal laws, rules, and regulations.

Currently, all fifty states have authorization to manage at least a portion of the RCRA program. One feature of the authorization process is interesting, in that changes to the federal regulations do not apply in authorized states until the authorized state changes its laws, rules, and regulations. See:

> *https://www.epa.gov/rcra/state-authorization-under-resource-conservation-and-recovery-act-rcra (accessed April 2025)*

for the current status of RCRA authorization nationwide.

That means that all of the regulatory changes described in this text become effective in different states at different times.

Comprehensive Environmental Response, Compensation, and Liability Act (CERCLA)

On December 11, 1980, in further reaction to Love Canal and other newly discovered hazardous waste cleanup sites, Congress also passed the "Superfund" law, officially known as the Comprehensive Environmental Response, Compensation, and Liability Act (CERCLA) – 42 U.S.C. §9601 et seq.

The Toxic Waste Act, or Superfund, is a federal law of the United States created to clean up deserted hazardous waste sites. The law authorizes the Environmental Protection Agency to investigate and analyze polluted sites and to order the responsible parties to clean up threatened areas or to clean the sites themselves in the absence of a responsible party.

This act provided money from industry and the federal government for the investigation and cleanup of hazardous waste sites where the responsible party could not be found or could not afford the work. CERCLA was funded by hazardous waste generation fees paid by industries.

Individual states also began to attach prorated fees to the generation and management of these wastes in order to help pay for the new program and to investigate and clean up the abandoned hazardous waste sites across the nation. There are two types of provisions in CERCLA with regard to the cleanup of a threatened site:

- *Removal actions* typically target localized releases that require immediate attention. Removal actions can be emergency, time critical, or non-time critical.
- *Remedy actions* target releases that may take longer to clean than removal actions. Remedial actions are permanent and minimize the risks involved. These actions can be performed only at EPA listed sites, the National Priorities List (NPL).

Cleaning up inactive hazardous waste sites is very different from managing hazardous waste generated from manufacturing or other active processes, mainly because at inactive or abandoned sites, the wastes are generally underground and, therefore, are difficult to access. Various hazardous waste cleanup technologies and remedy actions are discussed in more detail in Chapter 5.

Superfund was effective, but it was not funded well enough to tackle the rapidly growing number of hazardous waste cleanup sites that were being discovered. In October 1986, the Superfund Amendments and Reauthorization Act (SARA) was enacted. It amended CERCLA and increased the spectrum of Superfund. These broad changes also increased the fund to $9.3 billion [Bagai].

RCRA Reformed

Although the EPA and the state environmental agencies were making progress in regulating hazardous waste management with RCRA, Congress became concerned that too many hazardous wastes were being buried in landfills without treatment or stabilization, and some hazardous wastes were being improperly recycled, creating potential pollution problems for the future. To remedy this, RCRA was amended in 1984 by the Hazardous and Solid Waste Amendments (HSWA) (P.L. 98-616, 98 Stat. 3221), adding requirements for the USEPA to develop regulations that would encourage alternative waste management in order to reduce the use of landfills. This legislation resulted in the Land Disposal Restrictions program (LDR).

The LDR began to prescribe the manner in which industries must treat most hazardous wastes before the residuals could be land disposed. The USEPA also promulgated regulations allowing delegation of regulatory authority for much of the federal program to the fifty states, provided they develop laws, rules, and regulations that were at least as stringent as the federal program. This further added to industries' challenges, because each state's regulatory structure is somewhat different from the others.

The RCRA program is very complex, and the regulations are written in a confusing manner, with many references to other sections and subsections of the regulations. There are also many more cross-references to other sections, subsections, and sub-subsections, resulting in a maze of words almost impossible for facilities to attempt to comprehend in order to comply. To complicate matters further, individual states that have applied for and been granted authorization to run their own hazardous waste programs have, in many cases, developed policies that are more stringent than the federal regulations. This means an industry can do everything necessary to meet federal regulations and still be in violation of the hazardous waste regulations of an individual state or states.

Although the USEPA has published preambles to their regulations and explanations of parts of their rules and regulations, they have not yet published clear and comprehensive guidance in any one document so that industries might comply with all of its environmental rules and regulations. If we compare the USEPA with another federal agency, the USEPA does not provide guidelines nearly as comprehensive as those of the Internal Revenue Service (IRS), whose rules and regulations are even more voluminous, and arguably more complex than RCRA. The IRS provides forms for every tax for all potential regulated parties (filers). The IRS also provides detailed guidelines for completing these forms accurately. There are numerous professional accountants, tax preparers, and attorneys available to provide tax guidance and to assist in the filing of proper forms. The IRS guidelines have been written to help people and businesses know how to pay their taxes and how much to pay. Obviously, the U.S. government has a great deal of interest in making sure everyone pays their taxes accurately and on time, because these taxes are needed to fund most government programs. This may be the most critical difference between the tax laws and the environmental laws: The federal tax laws are necessarily explained in understandable terms for individuals and corporations to know how and how much to pay, while the environmental rules prioritize the protection of the nation's natural resources. Although the environmental laws generate fees as a way to pay for cleanups, encouraging compliance appears to be the higher priority.

1.5 PRINCIPLES OF HAZARDOUS WASTE MANAGEMENT

A major challenge facing engineers and regulatory staff is to develop methods to properly manage hazardous waste. As mentioned earlier, the first solid waste laws were written to protect human health and the environment by making sure every waste was safely disposed. In 1965, most industries and regulators agreed that the only "safe" disposal of wastes was to place them in land disposal units. This philosophy continued to be applied to hazardous waste disposal after RCRA passed in 1976, especially as it was generally much more economical to bury wastes than to treat them.

Hierarchy of Hazardous Waste Management

In 1990, the EPA, various state agencies, and industry organizations and companies recognized that disposing of waste in landfills should not be the first line of defense for protecting the environment. They all agreed that waste minimization and pollution prevention should be a higher priority. One popular hierarchy of waste is source reduction, recycling/reuse, treatment, and land disposal (only after complete treatment or stabilization of the wastes). The hazardous waste management hierarchy was developed to reduce the amount of hazardous waste generated, encourage the recycling and reuse of wastes where possible, and finally, to require the stabilization, neutralization, or destruction of the waste before placing the residual wastes in a lined landfill.

- *Source Reduction* is a combination of avoiding waste generation, generating the least amount and concentration, and generating the least toxic waste possible. Source reduction methods include:
  - Process modification (changing reaction temperatures, mixing times and method, and residence time intervals)
  - Product or material elimination (evaluating if products or materials are necessary, with customer input)
  - Control and management of inventory (not making products that won't be used)
  - Material substitution (using less toxic or less wasteful materials in production)
  - Improved housekeeping (eliminating wastes through less spillage and keeping product lines cleaner)
  - Return of unused material to suppliers (ideally for reuse or remanufacture)

- *Recycling/Reuse* includes using constituents of a waste material to manufacture a product, removing contaminants from a waste for reuse, and using a waste as a substitute product for a commercial product. Methods include:
    - Use as a fuel (cement production, lightweight aggregate, heat for processes)
    - Reuse of byproducts (substitution for raw materials)
    - Reprocessing (running bad or used products back through a process)
    - Reclamation (processing products and byproducts to produce products)
- *Treatment* is a process that modifies the physical, chemical, or biological character of a waste. Methods of treatment include processes such as: incineration, chemical treatment, biological treatment, thermal treatment, chemical stabilization, evaporation, and oxidation. These are discussed in Chapter 4.
- *Disposal* is the discharge, deposition, injection, dumping, spilling, leaking, or placing of any waste into the environment, including land, water, or air. Examples include: landfills, wastewater discharges, underground injection, and air releases. The waste management hierarchy directed that the first three options in the hierarchy must be utilized first, and only treated residuals should be disposed.

Land Disposal Restrictions Drive Hazardous Waste Treatment

To implement this hierarchy, Congress directed that only treated residuals could be land disposed, and to ensure that adequate treatment occurred (the third option in the hierarchy) before such waste was placed in landfills, they directed the USEPA to develop special rules called the Land Disposal Restrictions (LDRs) for hazardous waste generators to follow. These LDRs applied from the point the waste was generated until the treated residuals were placed in a landfill or beneficially reused [EPA 05].

Congress required the USEPA to establish treatment standards for different waste categories.

When the EPA was finished, they had prescribed three possible alternative standards for treatment, depending on the wastes:

- Required concentration levels
- Specified technologies
- A "no land disposal" designation for certain wastes

Hazardous waste management, including treatment technologies, is further discussed in Chapter 4.

NOTE

The Land Disposal Restrictions developed by the USEPA do not address the first two mandates in the hierarchy (source reduction and recycling/reuse) and therefore do not completely enforce the complete hazardous waste management hierarchy. The LDRs were not written to encourage facilities to reduce the amount of hazardous waste generated; rather the LDRs were written to require industries in the United States to treat or stabilize hazardous wastes before the residuals could be land disposed in a surface impoundment, waste pile, injection well, land treatment facility, salt bed formation, underground mine or cave, or concrete bunker or vault.

Because the LDRs drive the treatment of most hazardous wastes by prescribing either the levels of treatment or the required technology to be used by the facility managing the hazardous waste, they do not directly drive the upper parts of the hazardous waste management hierarchy such as source reduction, waste minimization, or recycling; however, the facilities generating the wastes can reduce their costs if they can find ways to reduce, reuse, or recycle the waste stream prior to treatment and disposal.

Determining Whether a Facility is Regulated Under RCRA

As discussed earlier, a facility in the United States must first determine if hazardous waste is generated or managed on the property to see if it is regulated as a hazardous waste. Hazardous waste is a part or subset of solid waste under RCRA, so if a material does not meet the definition of solid waste, it is not a hazardous waste.

Solid Waste

Recall from Section 1.1 that Congress defined solid waste as "any garbage, refuse, sludge from a wastewater treatment plant, water supply treatment plant, or air pollution control facility, and other discarded material, including solid, liquid, semisolid, or contained gaseous material resulting from industrial, commercial, mining, and from agricultural operations, and from community activities."

Also recall that the word "solid" in solid waste doesn't necessarily mean the solid phase of a material, in that the waste can be solid, liquid, or gaseous. This is confusing, but the law was written to regulate any problematic wastes that were not necessarily covered by any other environmental program.

For an example of the complex regulatory language in RCRA, let us look at 40 CFR Part 261, where solid waste is defined by the EPA as:

> "any discarded material that is not excluded under §261.4(a) or that is not excluded by a variance granted under §§260.30 and 260.31 or that is not excluded by a non-waste determination under §§260.30 and 260.34.
>
> (2)(i) A discarded material is any material which is:
> (A) Abandoned, as explained in paragraph (b) of this section; or
> (B) Recycled, as explained in paragraph (c) of this section; or
> (C) Considered inherently waste-like, as explained in paragraph (d) of this section; or
> (D) A military munition identified as a solid waste in §266.202" [EPA 11a].

Changes to the Definition of Solid Waste (DSW)

Under RCRA, hazardous wastes are a subset of solid wastes. A material must first be determined to be a solid waste; at which time a hazardous waste determination can be made. The federal solid waste program promulgated substantial revisions to the Solid Waste Rules, particularly aimed at the definition of solid waste, and tightening the rules concerning recycling of wastes, including hazardous waste. An excerpt from the USEPA fact sheet on these rules changes is included on the next few pages, with the portions concerning hazardous waste underlined.

On December 10, 2014, the Administrator signed the final revisions to the Definition of Solid Waste Rule, also known as the DSW rule. This represents a major environmental justice milestone by directly addressing impacts to communities, disproportionately borne by minority and low-income populations, from the mismanagement of hazardous materials sent to recycling, that is a central part of the final rule.

EPA conducted a rigorous environmental justice analysis of the DSW rule that examined the location of recycling facilities and their proximity and potential impact to adjacent residents.

The methodology and scope were developed through a broad public engagement and expert peer review process. The analysis identified significant regulatory gaps in the 2008 DSW rule, which could negatively impact communities adjacent to third party recyclers, including disproportionately impacting minority and low-income populations. The analysis will be available in the docket for the final rule once the rule is published.

In particular, EPA identified mismanagement by third-party hazardous materials recyclers as posing a risk of fires, explosions, accidents and releases of hazardous constituents to the environment. This is because the economics of commercial recycling contain market disincentives that encourage over-accumulation and mismanagement of hazardous secondary material. The 2008 DSW rule lacked the tools needed for proper oversight of these facilities by EPA, states and the communities affected by them.

The 2015 DSW rule addresses the market disincentives that encourage over-accumulation and mismanagement of hazardous secondary material in a way that helps encourage safe and legitimate recycling while addressing the need to protect communities. The DSW revisions provide a strong protection against the potential for mismanagement of hazardous materials intended for recycling, while allowing legitimate recycling activities to continue.

Specifically, the 2015 DSW rule withdraws the transfer-based exclusion and replaces it with the verified recycler exclusion.

This new provision requires that all hazardous materials recyclers operating under this provision have RCRA permits that address the materials or obtain a variance prior to operating under the exclusion. This will allow EPA and the states to use the RCRA permitting process or solid waste variance process to verify that a facility has established rigorous safety measures to manage the material.

Under the variance process, EPA and the states will also be able to review and approve the facilities' financial assurance plans to ensure that they're financially stable and that there will be funds available should the unexpected happen.

Requires that all entities subject to the new exclusions—both generators and recyclers—meet emergency response and preparedness requirements. This includes requiring facilities to make arrangements with local emergency response officials, which provides local fire departments and hospitals with critical information to enable them to tailor their preparations and response, thereby reducing risk to communities.

Includes a public participation requirement for recyclers seeking a verified recycler variance, so that communities are notified prior to recycling operations beginning and have a chance to weigh in on the environmental decisions that affect them.

Requires facilities seeking a verified recycler variance to address whether their activities pose a risk to nearby communities and whether their activities add to the cumulative environmental impacts.

For generators, the conditions of the exclusion include: (1) notifying the authorized state or EPA; (2) ensuring that hazardous secondary materials are contained; (3) maintaining records of shipments and confirmations of receipt of or transfers of hazardous secondary materials off-site; and (4) compliance with emergency preparedness requirements, which would be tailored according to the amount of hazardous secondary materials accumulated on-site. (These are similar to emergency preparedness requirements for hazardous waste generators.) The prohibition of sham recycling would also apply. Additionally, generators would no longer have to conduct an environmental audit of the recycler (since the recycler will have been verified by the state or EPA).

For reclaimers without a RCRA permit, in order to obtain a variance and become verified, the third-party reclaimer must address criteria that essentially mirrors the criteria under the reasonable efforts condition in the transfer-based exclusion. This includes: (1) demonstrate their recycling is legitimate, (2) must have financial assurance in place to properly manage the hazardous secondary material, (3) must not have had any formal enforcement actions for RCRA violations in the previous three years and is not classified as a significant non-complier with RCRA Subtitle C, or must provide credible evidence that the facility will manage the hazardous secondary materials properly, (4) must have the proper equipment, trained personnel, and meet emergency preparedness and response requirements to safely reclaim the material, (5) must manage the residuals from reclamation properly, and (6) must address risk to nearby communities from potential releases of the hazardous secondary material and in consideration of existing environmental stressors.

A recycler having a RCRA permit will have gone through the RCRA public participation process and have all of the protections currently required for recyclers managing hazardous waste. This includes financial assurance, emergency preparedness and response requirements, specific management standards (e.g., tank and container standards), and recordkeeping and reporting requirements.

These requirements are responsive to the concerns expressed by communities, and addresses the evidence of mismanagement by some third party recyclers documented in EPA's record.

The requirements provide strong protections against the potential for mismanagement of hazardous secondary materials intended for recycling.

The final rule also establishes a clear, uniform legitimate recycling standard for all hazardous secondary materials recycling that will improve compliance and help ensure that the hazardous secondary materials are in fact legitimately recycled, rather than illegally disposed.

Setting the standard for legitimate recycling in the regulations makes it substantially harder for facilities who are illegally disposing under the guise of recycling to continue to operate in the marketplace.

The four mandatory factors include: (1) the hazardous secondary material must provide a useful contribution to the recycling process or product; (2) the recycling process must produce a valuable product or intermediate; (3) the hazardous secondary material must be managed as a valuable commodity; and (4) the recycled product must be comparable to a legitimate product or intermediate.

The new DSW rule also includes several provisions that result in both resource conservation and economic benefits by encouraging certain types of in-process recycling and remanufacturing.

The rule affirms the legitimacy of the pre-2008 DSW exclusions, such as the scrap metal exclusion, and does not change the regulatory status of material legitimately recycled under these long-standing exclusions. (We are also not taking final action regarding notification and contained for existing pre-2008 exclusions based on a number of comments that questions our record support and encouraged us to do further study before making changes in order to avoid the potential for unintended consequences.)

The final rule re-affirms the legitimacy of in-process recycling and of commodity-grade recycled products, such as metal commodities. In both of these cases, the demonstration of a legitimate recycled product is straightforward. In-process recycling involves hazardous secondary materials being returned to the original production process. Additionally, recycled products that meet widely recognized commodity specifications are clearly commodities in wide use in commerce.

In both of these types of recycling, the recycled product is assumed to be legitimate.

The rule retains the exclusion for recycling under the control of the generator, including recycling onsite, within the same company and through toll manufacturing agreements. Generators who maintain control over their hazardous secondary materials and recycling processes have strong economic incentives to maintain oversight of, and responsibility for, the hazardous secondary material that is reclaimed.

Lastly, the final rule includes a targeted remanufacturing exclusion for certain higher-value hazardous spent solvents, which are being remanufactured into commercial-grade products. Facilities that recycle these solvents have strong economic incentives to properly manage these materials due to the profits realized from sale or use of the remanufactured solvents, rather than from fees charged for receiving the spent solvents. This exclusion allows manufacturers to reduce the use of virgin solvents, resulting in both economic and environmental benefits, including energy conservation and reduced greenhouse gas emissions.

Most states are authorized to administer and enforce the RCRA program in lieu of the federal program and these states will have to adopt the DSW rule before it becomes effective there. The impacts of the rule are dependent on how many states adopt the rule. Because the 2015 DSW rule addresses many of the concerns states raised about the 2008 DSW rule, state adoption rates—and thus cost savings—for the 2015 DSW rule may be much higher than the 2008 DSW final rule. EPA estimates that if 31 states and territories adopt the 2015 DSW rule, the rule will have an annual regulatory cost savings of $24 million as compared to baseline cost savings in the 8 states and territories that have adopted the 2008 DSW final rule.

More information about this rulemaking can be found on EPA's website at:

http://www.epa.gov/epawaste/hazard/dsw/rulemaking.htm."

In summary, this new rule modifies EPA's 2008 Definition of Solid Waste in response to communities' concerns about the mismanagement of hazardous waste secondary materials intended for recycling while ensuring responsible recycling. It also sets more stringent recycling standard for all hazardous secondary materials recycling.

Applying the Code

In translation, this language is subject to many interpretations. Many previous waste determinations are site specific and are not listed in any one place. Even if a facility hires experts to make solid waste and hazardous waste determinations, the experts' interpretations might not agree with the USEPA or state determination if the facility is inspected. To further complicate matters, individual states have different determinations of waste streams that are otherwise identical to those in other states.

Let's deal with this language piece by piece, addressing the essence of the referenced paragraphs, to see if we can make sense of it.

(A) *Abandoned* means materials for which the facility has no further use, and throws away, abandons, or destroys, or intends to do so. Examples include: products not used for their intended purpose; garbage, refuse, and sludge; process wastes; wastewater; obsolete chemicals; spent materials; process wastes; and might include products that will be recycled. For this reason, facilities must be careful not to keep chemicals that are outdated or no longer useful, because they can be considered abandoned, too.

(B) *Recycled* is a problematic term in the hazardous waste regulations, because the EPA traditionally allows each state authorized to implement RCRA to make the final determination on the applicability of RCRA on a recycling activity based on each particular facility and its processes. The regulations clarify whether RCRA regulations apply to waste materials that are recycled rather than disposed. The 1980 RCRA regulations exempted most materials destined for recycling from the definition of solid waste, so they were exempt from RCRA. The EPA changed that in 1985, on the heels of the discovery of a few "sham" recycling facilities.

Case Study

Sham Recycling

Recycling has been a controversial topic among hazardous waste inspectors since the beginning of the hazardous waste program. In the late 1970s, after RCRA was passed, many companies claimed to be recycling materials, but were, in fact, collecting money for taking the wastes and storing the materials in huge piles or burying them. Other facilities were burning hazardous wastes with little or no heating value and claiming they were burning the waste for energy recovery. Still others were actually performing recycling activities, but using such poor practices that their sites became contaminated to a point where the recycler couldn't afford the cleanup. Facilities should not make a claim of recycling without first clearing it with their home state environmental regulatory authority. Companies using these recyclers would be wise to check with the state regulatory authority where the facility is located to make sure the facility is properly permitted and, if necessary, should visit the site where the recycling happens to make sure it is legitimate. The penalties for illegitimate "sham" recycling are substantial and could lead to criminal charges.

Because each state has the final say on whether a particular facility is recycling or not, it is very wise to contact the state in which the facility is located to obtain a final regulatory answer.

! TIP

The customers (hazardous waste generators) of recyclers must be cautious about sending their waste for recycling, because if a recycler is not legitimate and a facility becomes a cleanup site, the government will look to all companies who shipped waste to cleanup sites as potentially responsible parties, all of whom are potentially liable for the cost of the cleanup. It is wise for companies to consult with their home state regulatory agency for interpretation on which activities are considered recycling and which are not.

(C) *Inherently waste-like* are materials that are considered too hazardous to be recycled, such as dioxin wastes (dioxin was a chemical found at Love Canal) and halogen-containing materials, such as hydrochloric acid (HCl), that are burned in halogen-acid furnaces. These wastes are considered highly toxic and require special handling.

(D) *Military munitions* are materials that are abandoned or removed from storage for disposal, burned, detonated, incinerated, or treated prior to disposal; or the munition is deteriorated or damaged to the point that it cannot be put into serviceable condition, and cannot reasonably be recycled or used for other purposes; or the munition has been declared a solid waste by an authorized military official.

The following case study involves an inspector who reluctantly has to ask a military installation how and where they disposed of their waste munitions.

Case Study

Military Munitions

Military munitions can be a touchy subject, because there are potential national security issues surrounding details revealed by military installations concerning the number of munitions they have on site and how many are disposed. An inspector at a military installation asked what they did with their expired military ordinance. The facility had armed guards twenty-four hours per day, seven days per week, 365 days per year, and the issue of unused, expired ammunition was raised, because it is a D001 hazardous waste. This question was initially met with some consternation, but knowing that there were armed guards and a firing range on site (see Figure 1.4), the inspector pressed for an answer.

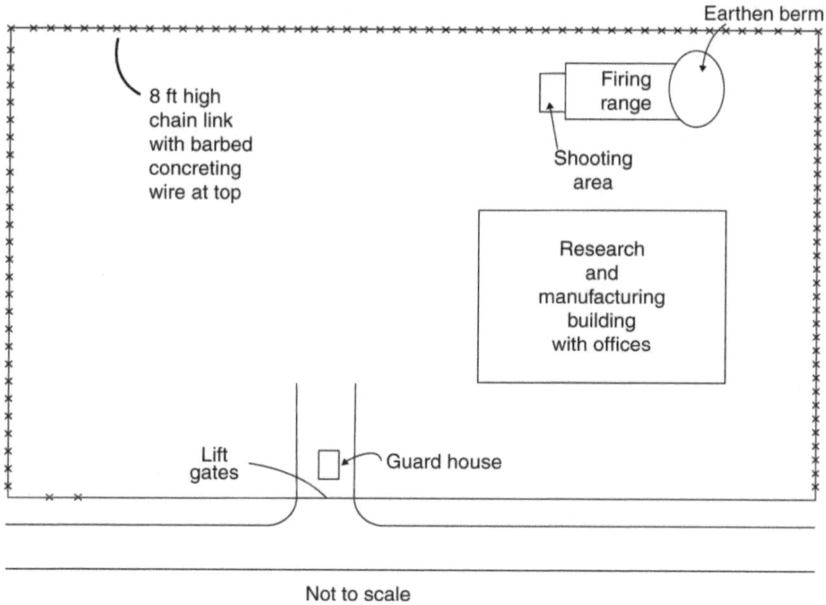

FIGURE 1.4 Military munitions (drawing by author).

Upon detailed questioning, the military installation revealed how they disposed of their munitions, and it became evident that the method of disposal was legal.

It is highly recommended that facilities contact staff at the USEPA and in the state where they are located to obtain their assistance in making hazardous waste determinations, or to have them make sure a waste determination is correct.

! TIP

It is unwise for facilities to make independent waste determinations without verifying the accuracy with their state officials, as they may find themselves in violation after an inspection. It is much worse for facilities to purposely not make waste determinations, because failure to make a waste determination is a major violation if the facility actually generates a hazardous waste, which could result in substantial fines. Requests for waste determination assistance from the USEPA or home state agencies can be made anonymously, and the regulatory staff is obliged to give an answer. If the facility wants a written record of the determination (highly recommended), it needs to reveal its identity.

As mentioned in the Introduction, the hazardous waste regulations, particularly the LDRs, direct the treatment standards that must be met for treatment or other forms of hazardous waste management before land disposal. Some of these standards even dictate the use of certain treatment technologies. Owners and operators of industries must understand these regulations well enough to meet the requirements. This means they need to properly identify their wastes, determine their regulatory status, choose the proper treatment standard for each waste stream, file required reports, retain certain necessary forms, and be prepared for possible (usually unannounced) compliance inspections. Experiencing a hazardous waste compliance inspection can be very daunting and nerve wracking for owners/operators who are new to the process, particularly because there is very little comprehensive published guidance on how to comply with this complex regulatory program.

Household Hazardous Waste

Household hazardous wastes are wastes, generated by homeowners, that would be considered hazardous wastes if they were generated by an industry. The United States Environmental Protection Agency (USEPA) excluded hazardous wastes generated in households (Household Hazardous Waste or HHW) from regulation as hazardous wastes, and all fifty states followed suit.

NOTE

The term HHW is a regulatory oxymoron in that, while the wastes indeed have all of the characteristics of a regulated hazardous waste, HHW is not regulated as a hazardous waste by the USEPA or by the states.

Even though HHW is not regulated as industrial hazardous waste, it is still regulated as a solid waste. Waste haulers and transfer facilities have the option to refuse to accept the more dangerous forms of these wastes into their collection trucks, transfer stations, incinerators, or landfills.

Rejection of a HHW by a garbage hauler forces a homeowner to find community cleanup days or to pay for much more expensive hazardous waste disposal methods. Properly disposing of household hazardous wastes can be very expensive. Improper or unsafe handling of these wastes can be very dangerous to the homeowner and to the waste management workers and facility.

The HHW exclusion applies only if the homeowner is the generator of the waste at their home. Wastes generated by contractors who renovate homes or companies who offer their services to clean out residences are not exempt from the hazardous waste regulations. These companies must make hazardous waste determinations on all of the wastes generated from renovation or demolition and must dispose of them legally under the federal and state laws, rules, and regulations. Contractors who renovate homes or offer their services to clean materials out of homes may also take advantage of the household hazardous waste exclusion but may (and often do) run into trouble with disposal facilities refusing to accept any hazardous materials. It is wise for the contractors to determine if any of the wastes are hazardous and dispose of them properly.

Likewise, communities or businesses who accept HHW at a cleanup day, where several homeowners bring their wastes, are required to properly pack and ship the wastes accompanied by a hazardous waste manifest and a LDR form.

Even though the USEPA allows hazardous wastes generated in households by homeowners to be thrown away with household trash, this practice can be detrimental to the environment. These wastes are dangerous for waste workers to handle, and often are toxic to the useful bacteria and enzymes that naturally break down the wastes in our sanitary landfills. These household wastes can also break out of their containers during handling and can expose workers at transfer stations, incinerators, or landfills to toxic fumes or chemical burns.

Side Note

Perhaps the most disturbing fact about HHW is that many homeowners do not recognize the health and safety issues associated with mishandling or misuse of these products. Homeowners are not trained as hazardous waste professionals, and often do not read the warning labels, wear the recommended protective clothing, nor provide adequate ventilation while using volatile products.

The misuse or lack of awareness of HHW by homeowners can be very dangerous and can cause injury, especially if certain cleaning chemicals are mixed, such as household bleach and ammonia, which produces chlorine gas.

The following case study is an example of a very dangerous and illegal misuse of a pesticide that should have been disposed as a HHW:

Case Study

Pesticide Application Leads to Hazardous Waste Generation and Cleanup

When one thinks of hazardous waste, one assumes that it is usually generated by an industry in the process of manufacturing a product. That is not always the case, as illustrated by this very unfortunate series of events.

In an apartment building in upstate New York in the early 1990s, tenants complained to their landlord about an infestation of cockroaches. Rather than call a certified exterminator, the landlord decided to take matters into his own hands. He distributed a pesticide to each of his tenants and told them to spray their own apartments. Giving the pesticide to the tenants would have been proper and legal, except for three major issues that turned out to be very costly to the landlord:

1. The pesticide he supplied to the tenants was chlordane, a chlorinated organic pesticide that had been banned from use in the United States since 1988.
2. Chlordane was listed to be used only for outdoor applications, mostly for termites. An indoor application of a pesticide designed and labeled for outdoor use is a major violation.
3. The health effects of chlordane in humans are on the nervous system, the digestive system, and the liver. These effects were seen mostly in people who swallowed chlordane mixtures. Large amounts of chlordane taken by mouth can cause convulsions and even death [ATSDR 94].

Discussion: It is impossible to measure the health effects of the chlordane fumes on the tenants because they all moved out of the apartment building, but the original complaints concerned the noxious fumes, and the tenants' inability to stay in their apartments. Offering the banned pesticide for use by the tenants was a serious violation but having them apply it inside their apartments was a much bigger problem. The chlordane soaked into the countertops and floors, including carpeting, so every item contaminated with chlordane had to be removed and disposed as a hazardous waste.

If the landlord had supplied the tenants with a legal pesticide, labeled for indoor use on roaches, he would have been within the law. He could have disposed of the chlordane at a community cleanup day at little or no cost as well. Instead, he tried to dispose of the banned pesticide by having his tenants use it in their apartments, and the resulting costs were staggering.

1.6 THE FUKUSHIMA DISASTER

The following case study concerns the March 11, 2011, Fukushima disaster and some of the attendant hazardous waste/hazardous materials contamination issues.

Case Study

Fukushima Disaster

The catastrophic earthquake and resulting tsunami off the coast of Japan on March 11, 2011, caused extensive damage to the Daiichi nuclear power plants in Fukushima (see Figure 1.5). Large amounts of radioactive water, steam, and seawater were created and released as the recovery teams cooled down the reactors.

This tragedy is distantly related to incidents such as the Love Canal hazardous waste landfill discussed earlier, but has some distinct differences, some of which are discussed here for the sake of clarity. In the United States, material that is radioactive is considered hazardous material because of its potential adverse health effects and is regulated solely by the Nuclear Regulatory Commission (NRC). Radioactive waste mixed with hazardous waste is regulated both by the USEPA and the NRC. While some of the wastes created and released at the Daiichi reactors may have hazardous waste mixed with radioactive waste, the press reports indicate the contamination of concern is primarily radioactive.

The most profound difference between the Love Canal and Fukushima disasters is the speed at which releases from the Fukushima disaster could affect human health. The time between the Fukushima disaster and human exposure is much faster than the time it took for migration of contaminants underground at Love Canal. At Fukushima, clouds of radioactively contaminated steam and gases were released into the air, and these pollutants migrated away from the site, carried by wind currents. Radioactive contamination attributed to this disaster was detected across the Pacific Ocean. A *Daily Mail* article, "Radiation from Japan's crippled nuclear plant detected in MILK in two U.S. states" [MAIL 11], indicates that

FIGURE 1.5 Aerial view of release at nuclear reactors at Daiichi plant, Fukushima, Japan (from asianews.it).

low levels of radiation were found in cow's milk in the states of Washington and California less than two months after the incident. In contrast, the underground migration of contaminants from the Love Canal landfill spanned approximately twenty years, from 1953 when the landfill was closed, until the mid-1970s.

The ongoing cleanup at Fukushima is very different from that at Love Canal. At Love Canal, the contaminants were generally confined by a leachate collection system and groundwater cutoff walls. The damaged Daiichi reactors are being decommissioned, a process that is currently estimated to take thirty to forty years at a cost of tens of billions of dollars. See

https://en.wikipedia.org/wiki/Fukushima_nuclear_accident_cleanup (accessed April 2025).

The greatest similarity between the two disasters may be that the Fukushima disaster made international news, and could result in new laws, rules, and regulations calling for safer operation, better emergency contingency plans, and cleanup after such disasters at nuclear power plants.

1.7 THE PFAS/PFOS ISSUE

The following case study concerns the early discovery of adverse health effects of some of the PFAS and PFOS compounds in the author's home state (New York State) and attendant hazardous waste/hazardous materials contamination issues and actions by New York State. Most of this information was from the websites of the New York State Department of Environmental Conservation and the New York State Department of Health,

https://dec.ny.gov/environmental-protection/site-cleanup/pfas.

Case Study

While the proposed federal regulations mentioned in the introduction to this book are set to go into effect shortly, several states took more immediate action to deal with the immediate threat of contaminated drinking water. Per- and Polyfluoroalkyl Substances (PFAS) are a group of chemicals used to make fluoropolymer coatings and products that resist heat, oil, stains, grease, and water. Fluoropolymer coatings are blends of resins and lubricants used in products such as water-repellent clothing, furniture, adhesives, paint and varnish, food packaging, heat-resistant non-stick cooking surfaces and insulation of electrical wires. They were also used in the production of firefighting foam, which increased the issue of contamination of surfaces and water.

Some PFAS do not break down easily and persist for a long time in the environment, especially in water. Their toxicity and persistence in the environment means they are a potential danger to public health and the environment. Additional information on potential health impacts can be found in a 2016 NYS Department of Health (DOH) letter contained in a regulatory impact statement at

https://extapps.dec.ny.gov/docs/remediation_hudson_pdf/part597erris.pdf

and a DOH Fact Sheet: Perfluorooctanoic Acid (PFOA) in Drinking Water, Hoosick Falls

https://www.health.ny.gov/environmental/investigations/drinkingwaterresponse/docs/atsdr_pfas_factsheet.pdf.

See fact sheet about Reducing PFAS in Private Wells at

https://extapps.dec.ny.gov/docs/remediation_hudson_pdf/pfasprivatewellsfs.pdf.

Chemicals in this group include perfluorooctanoic acid (PFOA) and perfluorooctane sulfonic acid (PFOS).

While PFAS and PFOS compounds were discovered in the 1940s and put to use in the 1950s, the public health impacts of some of the salts of these chemicals were not discovered until over fifty years later.

In early 2016, New York, joined by Vermont and New Hampshire, urged the U.S. Environmental Protection Agency (EPA) to acknowledge that PFOA contamination is a national problem that requires consistent federal guidelines. On May 19, 2016, EPA issued a lifetime health advisory of 70 parts per trillion (PPT) for long-term exposure to PFOA and PFOS in drinking water.

In January 2016, New York became the first state in the nation to regulate PFOA as a hazardous substance followed by the regulation of PFOS in April 2016. The regulation requires the proper storage of the substances and limited releases to the environment and enabled the state to use its legal authority and resources of the State Superfund program to advance investigations and cleanups of impacted sites. The Final Rule for PFOA and PFOS became effective on March 3, 2017. By finalizing the regulations, the state has solidified its authority to hold polluters accountable whenever PFOA and PFOS contamination is found. The federal rule is going into effect shortly.

This is an example of how development of chemical compounds can result in unintended consequences, thus the need for caution in their use and in disposal of wastes from their production.

In this chapter, you learned the definitions of solid waste and hazardous waste, discovered the history of solid and hazardous waste management, learned the fundamental principles of hazardous waste management, and read case studies on sham recycling, military munitions, a pesticide application that led to a major cleanup, and the nuclear reactor disaster.

In the next chapter, you will learn about the primary responsibilities of generators, learn the categories of hazardous waste, discover the relaxed standards for universal wastes, and learn about mixed hazardous and radioactive wastes and special state hazardous wastes.

You will learn why it is important to thoroughly verify your waste determinations, and you will also read case studies on improper waste identification, a company leaving town without warning, and an electroplater who claims he has no hazardous waste.

EXERCISES

1. What is the definition of hazardous waste?
   a. What does "characteristic hazardous waste" mean? (What are the characteristics that can make a waste hazardous?)
   b. What does "listed hazardous waste" mean? (Give an explanation of K-List, P-List, and U-List, respectively.)
2. How did Congress define solid waste?
3. What is the relationship between hazardous waste and solid waste?
4. What events prompted Congress to contemplate laws governing municipal waste?
5. What was the first law written to control municipal waste disposal?
   a. When was it enacted?
6. What events prompted Congress to contemplate laws governing hazardous wastes?
7. What was the first law written to control hazardous wastes?
   a. When was it enacted?
8. What was the first law written to clean up abandoned hazardous waste sites?
   a. When was it enacted?
9. When was RCRA first amended and what was it called?
10. Why were the amendments written?
11. What are the Land Disposal Restrictions?
12. What is Household Hazardous Waste?
13. How are Household Hazardous Wastes regulated?
14. What does State Authorization mean in the RCRA program? (Answer: EPA grants permission for a state to administer its own RCRA program if the state sets up a RCRA program at least as stringent as the USEPA program and EPA approves it.)
15. How does state authorization affect companies in that state? (Answer: After authorization is granted, the state laws, rules, and regulations become the effective enforcement mechanism in that state.)

REFERENCES

[ATSDR 94] Federal Agency for Toxic Substances and Disease Registry, Public Health Statement for Chlordane, May 1994, CAS# 12789-03-6, available online at *https://wwwn.cdc.gov/TSP/ToxProfiles/ToxProfiles.aspx?id=355&tid=62* (accessed April 2025).

[Bagai] Bagai, Eric. Toxic waste act. eHow.com, online at *https://en.wikipedia.org/wiki/Toxic_waste* (accessed April 2025).

[Engelhaupt 08] Engelhaupt, Erika. Happy birthday, Love Canal. *Chemical and Engineering News,* November 17, 2008, 86:46, pp. 46–53, online at *http://pubs.acs.org/cen/government/86/8646gov2.html* (accessed April 2025).

[EPA 05] U.S. Environmental Protection Agency. RCRA Training Module: Solid Waste and Emergency Response (5305W). *Introduction to land disposal restrictions* EPA530-K-05-013, 40 CFR Part 268. September 2005, online at *https://www.epa.gov/rcra/resource-conservation-and-recovery-act-rcra-training-module-about-land-disposal-restrictions* (accessed April 2025).

[EPA 11a] U.S. Environmental Protection Agency. eDocket CFR Part 261.2, online at *http://edocket.access.gpo.gov/cfr_2007/julqtr/pdf/40cfr261.2.pdf* (accessed April 2025).

[GovInfo] Solid Waste Disposal Act. eHow.com, online at *https://www.govinfo.gov/content/pkg/COMPS-893/pdf/COMPS-893.pdf* (accessed April 2025).

[MAIL 11] *Dailymail.com.* 2011. Radiation from Japan's crippled nuclear plant detected in MILK in two U.S. states. March 31, online at *http://www.dailymail.co.uk/news/article-1371930/Japan-nuclear-crisis-Fukushima-radiation-detected-MILK-2-US-states.html* (accessed April 2025).

[SWDA 65] *The Solid Waste Disposal Act* (SWDA) (P.L. 89-272, 79 Stat. 992) (October 20, 1965).

[WIKI 11a] Wikipedia.com, Sanitation in ancient Rome, online at *http://en.wikipedia.org/wiki/Sanitation_in_ancient_Rome* (accessed April 2025).

[WIKI 11b] Wikipedia.com, Love Canal, online at *http://en.wikipedia.org/wiki/Love_Canal (accessed April 2025).*

[Zuesse 81] Zuesse, Eric. Love Canal the truth seeps out. *Reason*, February 1981, online at *http://reason.com/archives/1981/02/01/love-canal* (accessed April 2025).

CHAPTER 2

Identification of Hazardous Waste

In This Chapter

- The primary responsibilities of generators
- Categories of hazardous waste
- Relaxed standards for universal wastes
- Mixed hazardous and radioactive wastes
- Special individual state hazardous wastes
- Verify waste determinations
- Case studies

2.1 THE GENERATORS' PRIMARY RESPONSIBILITIES

It is solely the generator's responsibility to determine if it generates or manages solid waste, and it is solely the generator's responsibility to determine if any of those solid wastes are hazardous wastes. Waste determinations are critical to businesses that generate wastes, because the determinations categorize the generators and apply regulatory requirements specific to their businesses. Many waste vendors and brokers solicit work, advertising their services for full-service waste determinations and management. It is potentially dangerous for businesses to put their full trust in these vendors and brokers, because it is the business owner who is ultimately responsible for their determinations.

2.2 CATEGORIES OF HAZARDOUS WASTE

As mentioned early in Chapter 1, "A Brief History of Hazardous Waste," in the United States, hazardous wastes are divided into two general categories—wastes that are hazardous because of their characteristics and wastes that are hazardous because they are specifically listed in the USEPA regulations.

Characteristic Hazardous Wastes

1. *Ignitable (D001)*: Ignitable wastes are liquids having a flash point < 60 °C (140 °F) and compressed gases with a pressure over 1 atmosphere that are flammable or that support combustion. They can create fires under certain conditions or can spontaneously combust.

   Examples of D001 wastes are acetone, benzene, waste gasoline, alcohols, naptha, petroleum distillates, and other used solvents such as xylene.

NOTE *The waste can also be ignitable if it is not a liquid and is capable under standard temperature and pressure, of causing fire through friction, absorption of moisture or spontaneous chemical changes and, when ignited, burns so vigorously and persistently that it creates a hazard. Also, if the waste is an oxidizer per 49 CFR 173.127, it can be ignitable.*

2. *Corrosive (D002)*: Corrosive wastes are concentrated acids or bases (pH ≤ 2, or ≥12.5) or liquids that are capable of corroding metal containers, such as storage tanks, drums, and barrels. Also, if it is a liquid and dissolves steel at the rate of 0.25"/year, it is corrosive.

   Examples of D002 wastes are acid from lead/acid batteries, etching solutions from printing/photography, ammonium solutions, hydroxide solutions, acid or alkaline cleaning solutions, rust removers, battery acid, and caustic hot tanks waste.
3. *Reactive (D003)*: Reactive wastes are unstable and can cause explosions, toxic fumes, gases, or vapors when exposed to water, heat, or increased pressure. A waste is reactive if it reacts violently with water, forms potentially explosive mixtures with water, generates toxic gases when mixed with water, contains cyanides or sulfides that are released when exposed to acid or alkaline materials, or is explosive.

   Examples of reactive wastes are cyanide plating wastes, waste concentrated bleaches, pressurized aerosol cans, and metallic sodium and potassium.
4. *Toxic (D004–D043)*: Toxic wastes are harmful or fatal when ingested or absorbed through the skin. Toxicity is defined through a laboratory procedure called the Toxicity Characteristic Leaching Procedure (TCLP). The TCLP helps identify wastes likely to leach concentrations of contaminants that might be harmful to human health or the environment.

Examples of toxic wastes are painting wastes that contain toxic metal-based pigments and/or certain solvents, such as Methyl Ethyl Ketone (MEK); treated wood waste where the treatment was done with "penta," or pentachlorophenol; and oily wastes, such as used oil filters that exceed the levels for benzene and/or lead.

NOTE *Hazardous wastes do not always have only one waste code assigned to them. Some wastes may have many hazardous waste codes.*

The D-List

EPA created the D-list, and any waste containing contaminants greater than or equal to the regulatory level, as determined by the Toxicity Characteristic Leaching Procedure (TCLP), is regulated as a characteristic hazardous waste.

TABLE 2.1 D-List, Except for D001, D002, and D003, Defined Previously

EPA HW No.[1]	Contaminant	CAS No.[2]	Regulatory Level (mg/L)
D004	Arsenic	7440–38–2	5.0
D005	Barium	7440–39–3	100.0
D018	Benzene	71–43–2	0.5
D006	Cadmium	7440–43–9	1.0
D019	Carbon tetrachloride	56–23–5	0.5
D020	Chlordane	57–74–9	0.03
D021	Chlorobenzene	108–90–7	100.0
D022	Chloroform	67–66–3	6.0
D007	Chromium	7440–47–3	5.0
D023	o-Cresol	95–48–7	[4]200.0
D024	m-Cresol	108–39–4	[4]200.0
D025	p-Cresol	106–44–5	[4]200.0
D026	Cresol		[4]200.0
D016	2,4-D	94–75–7	10.0
D027	1,4-Dichlorobenzene	106–46–7	7.5
D028	1,2-Dichloroethane	107–06–2	0.5
D029	1,1-Dichloroethylene	75–35–4	0.7
D030	2,4-Dinitrotoluene	121–14–2	[3]0.13
D012	Endrin	72–20–8	0.02
D031	Heptachlor (and its epoxide)	76–44–8	0.008
D032	Hexachlorobenzene	118–74–1	[3]0.13
D033	Hexachlorobutadiene	87–68–3	0.5
D034	Hexachloroethane	67–72–1	3.0
D008	Lead	7439–92–1	5.0
D013	Lindane	58–89–9	0.4
D009	Mercury	7439–97–6	0.2
D014	Methoxychlor	72–43–5	10.0
D035	Methyl ethyl ketone	78–93–3	200.0
D036	Nitrobenzene	98–95–3	2.0
D037	Pentrachlorophenol	87–86–5	100.0
D038	Pyridine	110–86–1	[3]5.0

(Continued)

TABLE 2.1 Continued

EPA HW No.[1]	Contaminant	CAS No.[2]	Regulatory Level (mg/L)
D010	Selenium	7782–49–2	1.0
D011	Silver	7440–22–4	5.0
D039	Tetrachloroethylene	127–18–4	0.7
D015	Toxaphene	8001–35–2	0.5
D040	Trichloroethylene	79–01–6	0.5
D041	2,4,5-Trichlorophenol	95–95–4	400.0
D042	2,4,6-Trichlorophenol	88–06–2	2.0
D017	2,4,5-TP (Silvex)	93–72–1	1.0
D043	Vinyl chloride	75–01–4	0.2

Source: EPA 40 CFR Part 261.20.
[1]Hazardous waste number.
[2]Chemical abstracts service number.
[3]Quantitation limit is greater than the calculated regulatory level. The quantitation limit therefore becomes the regulatory level.
[4]If o-, m-, and p-Cresol concentrations cannot be differentiated, the total cresol (D026) concentration is used. The regulatory level of total cresol is 200 mg/l.

Listed Hazardous Wastes

Wastes that are specifically listed in the USEPA hazardous waste regulations include the F-list (wastes from common manufacturing and industrial processes), K-list (wastes from specific industries), and P- and U-lists (wastes from commercial chemical products).

The F-List

The F-List identifies wastes from industrial and manufacturing processes, such as solvents that have been used in cleaning or degreasing operations. F-list wastes are called “non-specific” because they occur in different industry sectors. The F-listed wastes are known as wastes from non-specific sources.

TABLE 2.2 Complete F-List

Industry and EPA hazardous waste No.	Hazardous waste	Hazard code
Generic		
F001	The following spent halogenated solvents used in degreasing: Tetrachloroethylene, trichloroethylene, methylene chloride, 1,1,1-trichloroethane, carbon tetrachloride, and chlorinated fluorocarbons; all spent solvent mixtures/blends used in degreasing containing, before use, a total of ten percent or more (by volume) of one or more of the previous halogenated solvents or those solvents listed in F002, F004, and F005; and still bottoms from the recovery of these spent solvents and spent solvent mixtures	(T)

Industry and EPA hazardous waste No.	Hazardous waste	Hazard code
Generic		
F002	The following spent halogenated solvents: Tetrachloroethylene, methylene chloride, trichloroethylene, 1,1,1-trichloroethane, chlorobenzene, 1,1,2-trichloro-1,2,2-trifluoroethane, ortho-dichlorobenzene, trichlorofluoromethane, and 1,1,2-trichloroethane; all spent solvent mixtures/blends containing, before use, a total of ten percent or more (by volume) of one or more of the previous halogenated solvents or those listed in F001, F004, or F005; and still bottoms from the recovery of these spent solvents and spent solvent mixtures	(T)
F003	The following spent non-halogenated solvents: Xylene, acetone, ethyl acetate, ethyl benzene, ethyl ether, methyl isobutyl ketone, n-butyl alcohol, cyclohexanone, and methanol; all spent solvent mixtures/blends containing, before use, only the previous spent non-halogenated solvents; and all spent solvent mixtures/blends containing, before use, one or more of the previous non-halogenated solvents, and, a total of ten percent or more (by volume) of one or more of those solvents listed in F001, F002, F004, and F005; and still bottoms from the recovery of these spent solvents and spent solvent mixtures	(I)*
F004	The following spent non-halogenated solvents: Cresols and cresylic acid, and nitrobenzene; all spent solvent mixtures/blends containing, before use, a total of ten percent or more (by volume) of one or more of the previous non-halogenated solvents or those solvents listed in F001, F002, and F005; and still bottoms from the recovery of these spent solvents and spent solvent mixtures	(T)
F005	The following spent non-halogenated solvents: Toluene, methyl ethyl ketone, carbon disulfide, isobutanol, pyridine, benzene, 2-ethoxyethanol, and 2-nitropropane; all spent solvent mixtures/blends containing, before use, a total of ten percent or more (by volume) of one or more of the previous non-halogenated solvents or those solvents listed in F001, F002, or F004; and still bottoms from the recovery of these spent solvents and spent solvent mixtures	(I,T)
F006	Wastewater treatment sludges from electroplating operations except from the following processes: (1) Sulfuric acid anodizing of aluminum; (2) tin plating on carbon steel; (3) zinc plating (segregated basis) on carbon steel; (4) aluminum or zinc-aluminum plating on carbon steel; (5) cleaning/stripping associated with tin, zinc, and aluminum plating on carbon steel; and (6) chemical etching and milling of aluminum	(T)
F007	Spent cyanide plating bath solutions from electroplating operations	(R, T)
F008	Plating bath residues from the bottom of plating baths from electroplating operations where cyanides are used in the process	(R, T)
F009	Spent stripping and cleaning bath solutions from electroplating operations where cyanides are used in the process	(R, T)
F010	Quenching bath residues from oil baths from metal heat treating operations where cyanides are used in the process	(R, T)
F011	Spent cyanide solutions from salt bath pot cleaning from metal heat treating operations	(R, T)

(Continued)

TABLE 2.2 Continued

Industry and EPA hazardous waste No.	Hazardous waste	Hazard code
Generic		
F012	Quenching wastewater treatment sludges from metal heat treating operations where cyanides are used in the process	(T)
F019	Wastewater treatment sludges from the chemical conversion coating of aluminum except from zirconium phosphating in aluminum can washing when such phosphating is an exclusive conversion coating process. Wastewater treatment sludges from the manufacturing of motor vehicles using a zinc phosphating process will not be subject to this listing at the point of generation if the wastes are not placed outside on the land prior to shipment to a landfill for disposal and are either: disposed in a Subtitle D municipal or industrial landfill unit that is equipped with a single clay liner and is permitted, licensed, or otherwise authorized by the state; or disposed in a landfill unit subject to, or otherwise meeting, the landfill requirements in §258.40, §264.301, or §265.301. For the purposes of this listing, motor vehicle manufacturing is defined in paragraph (b)(4)(i) of this section, and (b)(4)(ii) of this section describes the recordkeeping requirements for motor vehicle manufacturing facilities	(T)
F020	Wastes (except wastewater and spent carbon from hydrogen chloride purification) from the production or manufacturing use (as a reactant, chemical intermediate, or component in a formulating process) of tri- or tetrachlorophenol, or of intermediates used to produce their pesticide derivatives. (This listing does not include wastes from the production of Hexachlorophene from highly purified 2,4,5-trichlorophenol.)	(H)
F021	Wastes (except wastewater and spent carbon from hydrogen chloride purification) from the production or manufacturing use (as a reactant, chemical intermediate, or component in a formulating process) of pentachlorophenol, or of intermediates used to produce its derivatives	(H)
F022	Wastes (except wastewater and spent carbon from hydrogen chloride purification) from the manufacturing use (as a reactant, chemical intermediate, or component in a formulating process) of tetra-, penta-, or hexachlorobenzenes under alkaline conditions	(H)
F023	Wastes (except wastewater and spent carbon from hydrogen chloride purification) from the production of materials on equipment previously used for the production or manufacturing use (as a reactant, chemical intermediate, or component in a formulating process) of tri- and tetrachlorophenols. (This listing does not include wastes from equipment used only for the production or use of Hexachlorophene from highly purified 2,4,5-trichlorophenol.)	(H)
F024	Process wastes, including but not limited to, distillation residues, heavy ends, tars, and reactor clean-out wastes, from the production of certain chlorinated aliphatic hydrocarbons by free radical catalyzed processes. These chlorinated aliphatic hydrocarbons are those having carbon chain lengths ranging from one to and including five, with varying amounts and positions of chlorine substitution. (This listing does not include wastewaters, wastewater treatment sludges, spent catalysts, and wastes listed in §261.31 or §261.32.)	(T)

Industry and EPA hazardous waste No.	Hazardous waste	Hazard code
Generic		
F025	Condensed light ends, spent filters and filter aids, and spent desiccant wastes from the production of certain chlorinated aliphatic hydrocarbons, by free radical catalyzed processes. These chlorinated aliphatic hydrocarbons are those having carbon chain lengths ranging from one to and including five, with varying amounts and positions of chlorine substitution	(T)
F026	Wastes (except wastewater and spent carbon from hydrogen chloride purification) from the production of materials on equipment previously used for the manufacturing use (as a reactant, chemical intermediate, or component in a formulating process) of tetra-, penta-, or hexachlorobenzene under alkaline conditions	(H)
F027	Discarded unused formulations containing tri-, tetra-, or pentachlorophenol or discarded unused formulations containing compounds derived from these chlorophenols. (This listing does not include formulations containing Hexachlorophene sythesized from prepurified 2,4,5-trichlorophenol as the sole component.)	(H)
F028	Residues resulting from the incineration or thermal treatment of soil contaminated with EPA Hazardous Waste Nos. F020, F021, F022, F023, F026, and F027	(T)
F032	Wastewaters (except those that have not come into contact with process contaminants), process residuals, preservative drippage, and spent formulations from wood preserving processes generated at plants that currently use or have previously used chlorophenolic formulations (except potentially cross-contaminated wastes that have had the F032 waste code deleted in accordance with §261.35 of this chapter or potentially cross-contaminated wastes that are otherwise currently regulated as hazardous wastes (i.e., F034 or F035), and where the generator does not resume or initiate use of chlorophenolic formulations). This listing does not include K001 bottom sediment sludge from the treatment of wastewater from wood preserving processes that use creosote and/or pentachlorophenol	(T)
F034	Wastewaters (except those that have not come into contact with process contaminants), process residuals, preservative drippage, and spent formulations from wood preserving processes generated at plants that use creosote formulations. This listing does not include K001 bottom sediment sludge from the treatment of wastewater from wood preserving processes that use creosote and/or pentachlorophenol	(T)
F035	Wastewaters (except those that have not come into contact with process contaminants), process residuals, preservative drippage, and spent formulations from wood preserving processes generated at plants that use inorganic preservatives containing arsenic or chromium. This listing does not include K001 bottom sediment sludge from the treatment of wastewater from wood preserving processes that use creosote and/or pentachlorophenol	(T)

(Continued)

TABLE 2.2 Continued

Industry and EPA hazardous waste No.	Hazardous waste	Hazard code
Generic		
F037	Petroleum refinery primary oil/water/solids separation sludge—Any sludge generated from the gravitational separation of oil/water/solids during the storage or treatment of process wastewaters and oily cooling wastewaters from petroleum refineries. Such sludges include, but are not limited to, those generated in oil/water/solids separators; tanks and impoundments; ditches and other conveyances; sumps; and stormwater units receiving dry weather flow. Sludge generated in stormwater units that do not receive dry weather flow, sludges generated from non-contact once-through cooling waters segregated for treatment from other process or oily cooling waters, sludges generated in aggressive biological treatment units as defined in §261.31(b)(2) (including sludges generated in one or more additional units after wastewaters have been treated in aggressive biological treatment units) and K051 wastes are not included in this listing. This listing does include residuals generated from processing or recycling oil-bearing hazardous secondary materials excluded under §261.4(a)(12)(i), if those residuals are to be disposed of	(T)
F038	Petroleum refinery secondary (emulsified) oil/water/solids separation sludge—Any sludge and/or float generated from the physical and/or chemical separation of oil/water/solids in process wastewaters and oily cooling wastewaters from petroleum refineries. Such wastes include, but are not limited to, all sludges and floats generated in: induced air flotation (IAF) units, tanks and impoundments, and all sludges generated in DAF units. Sludges generated in stormwater units that do not receive dry weather flow, sludges generated from non-contact once-through cooling waters segregated for treatment from other process or oily cooling waters, sludges and floats generated in aggressive biological treatment units as defined in §261.31(b)(2) (including sludges and floats generated in one or more additional units after wastewaters have been treated in aggressive biological treatment units) and F037, K048, and K051 wastes are not included in this listing	(T)
F039	Leachate (liquids that have percolated through land disposed wastes) resulting from the disposal of more than one restricted waste classified as hazardous under subpart D of this part. (Leachate resulting from the disposal of one or more of the following EPA Hazardous Wastes and no other Hazardous Wastes retains its EPA Hazardous Waste Number(s): F020, F021, F022, F026, F027, and/or F028.)	(T)

Source: EPA 40 CFR Part 260.32.
(T) Toxic Waste
(H) Acute Hazardous Waste
(I) Ignitable Waste
(C) Corrosive Waste
(R) Reactive Waste
(E) Toxicity Characteristic Waste
*(I, T) should be used to specify mixtures that are ignitable and contain toxic constituents.

NOTE *The D-List and F-List are included in their entirety in this text because they are relatively short. The text includes only the first two pages of the K-, P-, and U-lists, because they are very lengthy. For complete copies of these lists, please refer to the companion files included with this book.*

The K-list

The K-list identifies wastes that come from industries that manufacture specific products, such as pesticides or plastics. A partial list is shown as an example.

The complete K-list is included in the companion files.

TABLE 2.3 Example Portion of K-List

Industry and EPA hazardous waste No.	Hazardous waste	Hazard code
Wood preservation: K001	Bottom sediment sludge from the treatment of wastewaters from wood preserving processes that use creosote and/or pentachlorophenol	(T)
Inorganic pigments:		
K002	Wastewater treatment sludge from the production of chrome yellow and orange pigments	(T)
K003	Wastewater treatment sludge from the production of molybdate orange pigments	(T)
K004	Wastewater treatment sludge from the production of zinc yellow pigments	(T)
K005	Wastewater treatment sludge from the production of chrome green pigments	(T)
K006	Wastewater treatment sludge from the production of chrome oxide green pigments (anhydrous and hydrated)	(T)
K007	Wastewater treatment sludge from the production of iron blue pigments	(T)
K008	Oven residue from the production of chrome oxide green pigments	(T)
Organic chemicals:		
K009	Distillation bottoms from the production of acetaldehyde from ethylene	(T)
K010	Distillation side cuts from the production of acetaldehyde from ethylene	(T)
K011	Bottom stream from the wastewater stripper in the production of acrylonitrile	(R, T)
K013	Bottom stream from the acetonitrile column in the production of acrylonitrile	(R, T)
K014	Bottoms from the acetonitrile purification column in the production of acrylonitrile	(T)
K015	Still bottoms from the distillation of benzyl chloride	(T)

(Continued)

TABLE 2.3 Continued

Industry and EPA hazardous waste No.	Hazardous waste	Hazard code
K016	Heavy ends or distillation residues from the production of carbon tetrachloride	(T)
K017	Heavy ends (still bottoms) from the purification column in the production of epichlorohydrin	(T)
K018	Heavy ends from the fractionation column in ethyl chloride production	(T)
K019	Heavy ends from the distillation of ethylene dichloride in ethylene dichloride production	(T)
K020	Heavy ends from the distillation of vinyl chloride in vinyl chloride monomer production	(T)
K021	Aqueous spent antimony catalyst waste from fluoromethanes production	(T)
K022	Distillation bottom tars from the production of phenol/acetone from cumene	(T)
K023	Distillation light ends from the production of phthalic anhydride from naphthalene	(T)
K024	Distillation bottoms from the production of phthalic anhydride from naphthalene	(T)
K025	Distillation bottoms from the production of nitrobenzene by the nitration of benzene	(T)
K026	Stripping still tails from the production of methy ethyl pyridines	(T)
K027	Centrifuge and distillation residues from toluene diisocyanate production	(R, T)
K028	Spent catalyst from the hydrochlorinator reactor in the production of 1,1,1-trichloroethane	(T)
K029	Waste from the product steam stripper in the production of 1,1,1-trichloroethane	(T)
K030	Column bottoms or heavy ends from the combined production of trichloroethylene and perchloroethylene	(T)
K083	Distillation bottoms from aniline production	(T)
K085	Distillation or fractionation column bottoms from the production of chlorobenzenes	(T)
K093	Distillation light ends from the production of phthalic anhydride from ortho-xylene	(T)
K094	Distillation bottoms from the production of phthalic anhydride from ortho-xylene	(T)
K095	Distillation bottoms from the production of 1,1,1-trichloroethane	(T)
K096	Heavy ends from the heavy ends column from the production of 1,1,1-trichloroethane	(T)
K103	Process residues from aniline extraction from the production of aniline	(T)

Industry and EPA hazardous waste No.	Hazardous waste	Hazard code
K104	Combined wastewater streams generated from nitrobenzene/aniline production	(T)
K105	Separated aqueous stream from the reactor product washing step in the production of chlorobenzenes	(T)
K107	Column bottoms from product separation from the production of 1,1-dimethylhydrazine (UDMH) from carboxylic acid hydrazines	(C,T)

Source: EPA 40 CFR §§260.20 and 260.22 and listed in appendix IX.
(T) Toxic Waste
(H) Acute Hazardous Waste
(I) Ignitable Waste
(C) Corrosive Waste
(R) Reactive Waste
(E) Toxicity Characteristic Waste
*(I, T) should be used to specify mixtures that are ignitable and contain toxic constituents.

The P- and U-lists

The P- and U-lists are unused commercial products, such as pesticides and pharmaceuticals that cannot be used or sold. These wastes can also be discarded commercial chemical products, off-specification species, container residues, and spill residues. Again, only a few sample pages are included in this text. The P-list items are "acutely hazardous" wastes, meaning they are more toxic than the others. The complete P- and U-Lists are included in the companion files.

TABLE 2.4 Example Portion of P-List

Hazardous waste No.	Chemical abstracts No.	Substance
P023	107–20–0	Acetaldehyde, chloro-
P002	591–08–2	Acetamide, N-(aminothioxomethyl)-
P057	640–19–7	Acetamide, 2-fluoro-
P058	62–74–8	Acetic acid, fluoro-, sodium salt
P002	591–08–2	1-Acetyl-2-thiourea
P003	107–02–8	Acrolein
P070	116–06–3	Aldicarb
P203	1646–88–4	Aldicarb sulfone
P004	309–00–2	Aldrin
P005	107–18–6	Allyl alcohol
P006	20859–73–8	Aluminum phosphide (R,T)
P007	2763–96–4	5-(Aminomethyl)-3-isoxazolol

(Continued)

TABLE 2.4 Continued

Hazardous waste No.	Chemical abstracts No.	Substance
P008	504–24–5	4-Aminopyridine
P009	131–74–8	Ammonium picrate (R)
P119	7803–55–6	Ammonium vanadate
P099	506–61–6	Argentate(1-), bis(cyano-C)-, potassium
P010	7778–39–4	Arsenic acid H_3AsO_4
P012	1327–53–3	Arsenic oxide As_2O_3
P011	1303–28–2	Arsenic oxide As_2O_5
P011	1303–28–2	Arsenic pentoxide
P012	1327–53–3	Arsenic trioxide
P038	692–42–2	Arsine, diethyl-
P036	696–28–6	Arsonous dichloride, phenyl-
P054	151–56–4	Aziridine
P067	75–55–8	Aziridine, 2-methyl-
P013	542–62–1	Barium cyanide
P024	106–47–8	Benzenamine, 4-chloro-
P077	100–01–6	Benzenamine, 4-nitro-
P028	100–44–7	Benzene, (chloromethyl)-
P042	51–43–4	1,2-Benzenediol, 4-[1-hydroxy-2-(methylamino)ethyl]-, (R)-
P046	122–09–8	Benzeneethanamine, alpha,alpha-dimethyl-
P014	108–98–5	Benzenethiol
P127	1563–66–2	7-Benzofuranol, 2,3-dihydro-2,2-dimethyl-, methylcarbamate
P188	57–64–7	Benzoic acid, 2-hydroxy-, compd. with (3aS-cis)-1,2,3,3a,8,8a-hexahydro-1,3a,8-trimethylpyrrolo[2,3-b]indol-5-yl methylcarbamate ester (1:1)
P001	[1]81–81–2	2H-1-Benzopyran-2-one, 4-hydroxy-3-(3-oxo-1-phenylbutyl)-, & salts, when present at concentrations greater than 0.3%
P028	100–44–7	Benzyl chloride
P015	7440–41–7	Beryllium powder
P017	598–31–2	Bromoacetone
P018	357–57–3	Brucine
P045	39196–18–4	2-Butanone, 3,3-dimethyl-1-(methylthio)-, O-[(methylamino)carbonyl] oxime
P021	592–01–8	Calcium cyanide
P021	592–01–8	Calcium cyanide $Ca(CN)^2$
P189	55285–14–8	Carbamic acid, [(dibutylamino)- thio]methyl-, 2,3-dihydro-2,2-dimethyl-7-benzofuranyl ester

Hazardous waste No.	Chemical abstracts No.	Substance
P191	644–64–4	Carbamic acid, dimethyl-, 1-[(dimethyl-amino)carbonyl]- 5-methyl-1H-pyrazol-3-yl ester
P192	119–38–0	Carbamic acid, dimethyl-, 3-methyl-1- (1-methylethyl)-1H- pyrazol-5-yl ester
P190	1129–41–5	Carbamic acid, methyl-, 3-methylphenyl ester
P127	1563–66–2	Carbofuran
P022	75–15–0	Carbon disulfide
P095	75–44–5	Carbonic dichloride
P189	55285–14–8	Carbosulfan
P023	107–20–0	Chloroacetaldehyde
P024	106–47–8	p-Chloroaniline

Source: EPA 40 CFR §§260.20 and 260.22 and appendix IX.
(T) Toxic Waste
(H) Acute Hazardous Waste
(I) Ignitable Waste
(C) Corrosive Waste
(R) Reactive Waste
(E) Toxicity Characteristic Waste
*(I, T) should be used to specify mixtures that are ignitable and contain toxic constituents.

TABLE 2.5 Example Portion of U-List

Hazardous waste No.	Chemical abstracts No.	Substance
U394	30558–43–1	A2213.
U001	75–07–0	Acetaldehyde (I)
U034	75–87–6	Acetaldehyde, trichloro-
U187	62–44–2	Acetamide, N-(4-ethoxyphenyl)-
U005	53–96–3	Acetamide, N-9H-fluoren-2-yl-
U240	194–75–7	Acetic acid, (2,4-dichlorophenoxy)-, salts & esters
U112	141–78–6	Acetic acid ethyl ester (I)
U144	301–04–2	Acetic acid, lead(2+) salt
U214	563–68–8	Acetic acid, thallium(1+) salt
see F027	93–76–5	Acetic acid, (2,4,5-trichlorophenoxy)-
U002	67–64–1	Acetone (I)
U003	75–05–8	Acetonitrile (I,T)
U004	98–86–2	Acetophenone
U005	53–96–3	2-Acetylaminofluorene

(Continued)

TABLE 2.5 Continued

Hazardous waste No.	Chemical abstracts No.	Substance
U006	75–36–5	Acetyl chloride (C,R,T)
U007	79–06–1	Acrylamide
U008	79–10–7	Acrylic acid (I)
U009	107–13–1	Acrylonitrile
U011	61–82–5	Amitrole
U012	62–53–3	Aniline (I,T)
U136	75–60–5	Arsinic acid, dimethyl-
U014	492–80–8	Auramine
U015	115–02–6	Azaserine
U010	50–07–7	Azirino[2′,3′:3,4]pyrrolo[1,2-a]indole-4,7-dione, 6-amino-8-[[(aminocarbonyl) oxy]methyl]-1,1a,2,8,8a,8b-hexahydro-8a-methoxy-5-methyl-, [1aS-(1aalpha, 8beta,8aalpha,8balpha)]-
U280	101–27–9	Barban
U278	22781–23–3	Bendiocarb
U364	22961–82–6	Bendiocarb phenol
U271	17804–35–2	Benomyl
U157	56–49–5	Benz[j]aceanthrylene, 1,2-dihydro-3-methyl-
U016	225–51–4	Benz[c]acridine
U017	98–87–3	Benzal chloride
U192	23950–58–5	Benzamide, 3,5-dichloro-N-(1,1-dimethyl-2-propynyl)-
U018	56–55–3	Benz[a]anthracene
U094	57–97–6	Benz[a]anthracene, 7,12-dimethyl-
U012	62–53–3	Benzenamine (I,T)
U014	492–80–8	Benzenamine, 4,4′-carbonimidoylbis[N,N-dimethyl-
U049	3165–93–3	Benzenamine, 4-chloro-2-methyl-, hydrochloride
U093	60–11–7	Benzenamine, N,N-dimethyl-4-(phenylazo)-
U328	95–53–4	Benzenamine, 2-methyl-
U353	106–49–0	Benzenamine, 4-methyl-
U158	101–14–4	Benzenamine, 4,4′-methylenebis[2-chloro-
U222	636–21–5	Benzenamine, 2-methyl-, hydrochloride
U181	99–55–8	Benzenamine, 2-methyl-5-nitro-
U019	71–43–2	Benzene (I,T)
U038	510–15–6	Benzeneacetic acid, 4-chloro-alpha-(4-chlorophenyl)-alpha-hydroxy-, ethyl ester

Hazardous waste No.	Chemical abstracts No.	Substance
U030	101–55–3	Benzene, 1-bromo-4-phenoxy-
U035	305–03–3	Benzenebutanoic acid, 4-[bis(2-chloroethyl)amino]-
U037	108–90–7	Benzene, chloro-
U221	25376–45–8	Benzenediamine, ar-methyl-
U028	117–81–7	1,2-Benzenedicarboxylic acid, bis(2-ethylhexyl) ester
U069	84–74–2	1,2-Benzenedicarboxylic acid, dibutyl ester
U088	84–66–2	1,2-Benzenedicarboxylic acid, diethyl ester
U102	131–11–3	1,2-Benzenedicarboxylic acid, dimethyl ester
U107	117–84–0	1,2-Benzenedicarboxylic acid, dioctyl ester

Source: EPA 40 CFR §§260.20 and 260.22 and appendix IX.
[1]CAS Number given for parent compound only.

2.2(A) NEW HAZARDOUS SUBSTANCES (PFAS/PFOS)

As mentioned in the Introduction, the USEPA has proposed to add nine particular per- and polyfluoroalkyl compounds, their salts, and their structural isomers, to its list of hazardous constituents in Title 40 of the Code of Federal Regulations Part 261 Appendix VIII. These nine PFAS are:

1. Perfluorooctanoic acid
2. Perfluorooctanesulfonic acid
3. Perfluorobutanesulfonic acid
4. Hexafluoropropylene oxide-dimer acid
5. Perfluorononanoic acid
6. Perfluorohexanesulfonic acid
7. Perfluorodecanoic acid
8. Perfluorohexanoic acid
9. Perfluorobutanoic acid

These formulations were discovered in the 1940s and touted as "miracle" type compounds that are repellent to oil, water, stains, and soil, resistant to chemical and temperature, resist friction, and exhibit surfactant properties. They started being used commercially in the 1950s.

According to the Interstate Technology Regulatory Council (ITRC), PFAS compounds alone have more than 200 uses for more than 1,400 individual PFAS.

According to the ITRC, most polymer PFAS are considered to pose less risk to human health and the environment than some nonpolymer PFAS.

Potential health concerns have been expressed for PFAS in cosmetics, face masks, fluorinated HDPE containers, food packaging, and pesticides, to name a few.

PFOS is a part of the class of PFAS and is used in many consumer and industrial products as well, including carpets, upholstered furniture, non-stick cookware, and firefighting foam.

Most of the PFAS and PFOS salts are water soluble and very difficult to remove during treatment.

The USEPA cited several potential health effects to exposure to PFAS compounds, reproductive effects, developmental delays in children, increased risk of some cancers, reduced immune system to fight infections, interference with natural hormones, and increased cholesterol levels.

The designation of these compounds as Hazardous Constituents is important, as the USEPA has proposed a rule that would broaden the definition of "hazardous waste" applicable to RCRA corrective action. The proposed rule would use EPA's interpretation that RCRA provides it authority to address and take corrective action for releases of any substances that meet the broad statutory definition of "hazardous waste." Under RCRA's Corrective Action Program, the EPA and authorized states manage investigations and cleanups of hazardous waste and hazardous constituents at treatment, storage and disposal facilities (TSDFs). The addition of PFAS and PFOS compounds to the list of hazardous constituents would make these compounds subject to corrective action nationwide. This could include areas where firefighting foam had been applied at airports and highways when the pavement is removed. The removed pavement would be a potentially hazardous waste, subject to all the hazardous waste rules and regulations, from identification to manifesting, to shipment, and disposal.

2.3 RELAXED STANDARDS FOR UNIVERSAL WASTES

Universal Wastes

The USEPA applied the term "Universal Waste" to certain types of wastes that are generated on a regular basis by almost every business and industry (universally). Many of these wastes are so commonly generated that their associated hazards and dangers are often downplayed or ignored.

The USEPA has set up streamlined rules and regulations to define these universal wastes and to encourage better collection and recycling:

- batteries
- pesticides
- mercury-containing equipment
  - thermostats
  - lamps

Businesses and industries enjoy relaxed rules and regulations, which offer many advantages from how they must manage regular non-acute hazardous waste, as follows:

- Labeling requirements are simplified—universal waste doesn't need to have a hazardous waste label (see Figure 2.1). Instead, the labels should read:
  - For batteries, "Universal Waste-Battery(ies)" or "Waste Battery(ies)" or "Used Battery(ies)"
  - For pesticides, "Universal Waste-Pesticide(s)" or "Waste Pesticide(s)"

   - For mercury switches, "Universal Waste-Mercury Switch(es)" or "Waste Mercury Switch(es)" or "Used Mercury Switch(es)"
   - For thermostats, "Universal Waste-Mercury Thermostat(s)" or "Waste Mercury Thermostat(s)" or "Used Mercury Thermostat(s)"
   - For lamps, "Universal Waste-Lamp(s)" or "Waste Lamp(s)" or "Used Mercury Lamp(s)"
- The container must be marked with accumulation start date. Storage on site for one year or longer is allowed if the extended storage time is solely to accumulate sufficient quantities necessary for proper recovery, treatment, or disposal. (In the experience of the Author, this seldom, if ever, occurs.) If it does occur, be careful to document and demonstrate the reason(s) for the extended storage.
- Universal Waste does not count toward hazardous waste generator status or level.
- There are no hazardous waste fees or taxes on Universal Wastes.
- Small Quantity Handlers of universal waste are not required to have an EPA generator identification number.
- Universal Waste handlers may accept universal waste from other generators.
- Universal Waste generators may use shipping papers or bills of lading rather than the uniform hazardous waste manifest.
- Small quantity handlers (those who accumulate less than 11,000 pounds at any one time) are not subject to record-keeping requirements.
- Universal Waste generators may self-transport using the less stringent universal waste transporter requirements.
- Universal Waste generators are not required to use licensed hazardous waste transporters.

FIGURE 2.1 Fiber drum with universal waste fluorescent lamps. (Photo courtesy of Faith Baptist Church, New York.)

It should be noted that U.S. Department of Transportation regulations might still apply to the material and mode of shipment.

NOTE

The universal waste rules apply only to the lists of products in their original containers or casings. If the original casing breaks—whether it is a switch, a battery, a thermostat, or a lamp—the waste must promptly be properly contained, cleaned up, identified, and managed as a hazardous waste.

Every company has the option to manage its universal waste as regular hazardous waste, following all of the rules, using hazardous waste manifests, and having those wastes count toward their annual hazardous waste generation. Some companies elect to do this because they don't generate sufficient universal waste to make it worthwhile to distinguish between the two types of waste.

! TIP

Some states have defined additional wastes as universal wastes. Check with your home state regulatory agency or their Web site to take full advantage of these relaxed rules.

2.4 MIXED WASTES

Mixed wastes are hazardous wastes that are mixed with low-level radioactively contaminated waste from industrial or research work (see Figure 2.2). They include items such as paper, rags, plastic bags, or water-treatment residues.

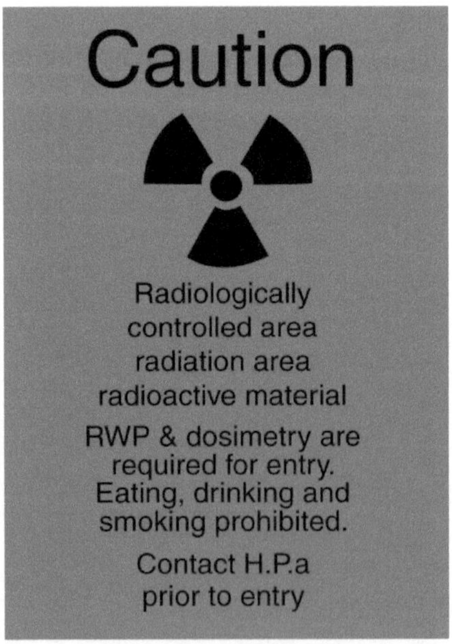

FIGURE 2.2 Radioactive symbol. (Image from Nuclear Regulatory Commission online at *https://www.nrc.gov/reading-rm/basic-ref/glossary/radiation-warning-symbol.html*.)

The regulation of mixed waste is much more complicated than hazardous waste. Mixed wastes are regulated jointly by the USEPA and the Nuclear Regulatory Commission (NRC), with the EPA regulating the hazardous waste portion of the waste and the NRC regulating the radiological portion. The Atomic Energy Act (AEA) and RCRA are in agreement for the most part, but whenever there is a conflict, the provisions of the AEA take precedence over RCRA.

One of the biggest problems with mixed waste is that there are often no treatment facilities available to manage the waste, so the waste either has to be treated on site in containers, to remove the hazardous portion, or be placed in long-term storage until either the radiological portion decays to a point where it is safe enough to treat the hazardous portion, or a viable treatment solution becomes available. A discussion of various types of storage facilities (salt dome formations, salt bed formations, underground mines, and underground caves) can be found in Chapter 4.

The best solution to resolving this waste management dilemma is to minimize the amounts and concentrations of the mixed waste generated to the least amounts and concentrations possible. This move up the hazardous waste management hierarchy can produce less mixed waste in the future.

2.5 SPECIAL INDIVIDUAL STATE HAZARDOUS WASTES

To complicate matters further, individual states are allowed to promulgate hazardous waste regulations that are more stringent than the federal rules and regulations. Some states identify additional substances as hazardous waste in their individual state. For example, New York State identifies polychlorinated biphenyls (PCBs) as hazardous waste. PCB wastes are not regulated as hazardous waste under RCRA. Instead, they are regulated by the federal Toxic Substances Control Act (TSCA).

The lamp ballasts shown in Figure 2.3 would be regulated as a hazardous waste in New York State if they contained PCBs at a concentration of 50 parts per million or higher.

FIGURE 2.3 Lamp ballasts containing PCBs. (Photo courtesy of Faith Baptist Church, New York.)

Some New England states identify waste oil as hazardous waste, and others do not. These differences make compliance with the hazardous waste rules difficult, particularly if the wastes are shipped from state to state or across several states for disposal. It is possible for wastes to go from hazardous to nonhazardous and back to hazardous as they cross the nation or are shipped into other countries. These regulatory differences from state to state can be very difficult to track, and transporters and waste management facilities often violate the rules, knowingly or unknowingly. For more details on how individual states differ from the USEPA in their waste identification, visit each state's individual Web sites or visit the USEPA Web site [EPA 11b].

! TIP *It is highly advisable for all facilities to contact the environmental regulatory staff in the state where waste is generated to see if there are special waste streams that are considered hazardous waste in that state.*

2.6 ALWAYS VERIFY WASTE IDENTIFICATION WITH HOME STATE REGULATORS

It is unwise for facilities to make independent solid and hazardous waste determinations for several reasons:

1. Failure to make a proper solid or hazardous waste determination is potentially a very serious violation, carrying possible monetary penalties.
2. It is critical to know if the facility must comply with solid and hazardous waste laws, rules, and regulations; and if it needs to comply, what exact compliance procedures must be followed.
3. Even if a facility makes a solid or hazardous waste determination in good faith, it is still possible to be contrary to the interpretation of the regulatory agencies and to face possible penalties. Having a written determination from a regulatory agency goes a long way toward convincing compliance staff that the facility tried its best to make a proper determination.

! TIP

Readers are advised to always ask questions of the agencies that regulate the wastes in question. The USEPA has staff that will provide regulatory interpretations, but it is preferred that the facility contact the appropriate state hazardous waste regulatory agency(ies) where they do business. If the facility doesn't want to reveal its identity, calls can be made anonymously, and the regulatory staff will still answer their questions.

Document All Waste Determinations

One of the first questions a hazardous waste inspector asks when he inspects a facility is, "Have you made waste determinations on all of the waste generated at the facility?" Make sure you have the determination(s) stored in an easily accessible, separate file so the inspector can review them. It is critical to document your waste determination(s) thoroughly, including conversations or other correspondence you have had from federal or state regulators.

! TIP

If a facility generates or manages hazardous wastes at a rate greater than 100 kilograms (220 pounds) per month, it must obtain a federal hazardous waste generator identification number (USEPA ID #). This number is generated by the USEPA regional offices, which are listed on the EPA Web site [EPA 11c]. These USEPA ID numbers are site specific. It is imperative to have a USEPA ID number for every physical location where hazardous waste is generated at a rate of more than 100 KG per month. Facilities generating less than 100 KG per month do not need a USEPA ID number.

2.7 CASE STUDIES

In the following case study, an owner/operator learns that improper waste identification can be costly, but for a surprising reason. The reader's first thought might be that improper identification is costly because regulators might discover that a company is not claiming its waste is hazardous. The following case study has an interesting twist, when one company learns that they are needlessly paying hazardous waste generation fees and disposal costs for wastes that are not hazardous.

Case Study

Improper Waste Identification Can be Costly

An inspector visited a company that designed and produced specialty mechanical machine parts. When the inspector introduced himself and explained his purpose, the company representative was hesitant and uncooperative, asking if the inspector had a search warrant. After the inspector explained his regulatory authority, and that if he needed a warrant he could get one quickly, the facility representative consented to the inspection. The company had relied on a waste management broker/transporter firm to do their hazardous waste identifications, and some of the determinations were not correct. Several of the wastes that were identified by the broker as being hazardous wastes were, in fact, not hazardous wastes, and a few other wastes that were identified as nonhazardous wastes were actually hazardous wastes.

Based on these erroneous determinations, the company listed itself as a Large Quantity Generator and was spending large amounts of money on generator annual reports, generator fees, and excess hazardous waste disposal costs, not to mention the extra time spent learning and complying with those rules. After analyzing the waste streams and interviewing the facility manager, the inspector informed the company that their generation rate made them only a Conditionally Exempt Small Quantity Generator (now called a Very Small Quantity Generator), with greatly reduced regulatory and financial requirements. This correction to the waste determinations saved the company thousands of dollars per year.

The inspector explained to the company representative that his company was ultimately responsible for making waste determinations, and that the waste broker/transporter that had given him incorrect information had no liability. When the inspector finished explaining the errors and the serious consequences of improper identification, he showed the facility representative how the incorrect waste determinations were costing him thousands of dollars every year, as the broker/transporter was charging his company hazardous waste prices for solid waste disposal. Because it was their first offense, the inspector issued the company a warning letter and gave them thirty days to correct the mistakes. The company immediately corrected the mistakes and fired their waste broker/transporter. The company representative was very grateful for the inspection and called the inspector several times after that to make sure he properly identified new waste streams. He told the inspector that if he had known how easy it was to get assistance on complying with the hazardous waste rules, he would never have relied on the waste broker/transporter for his waste management advice.

Regulatory offices take many anonymous calls from companies with hazardous waste questions and are generally pleased when they are called. These offices give answers that help reduce violations and make the inspectors' jobs easier and more pleasant.

In the next case study, we find out that companies are not legally obligated to notify anyone that they are going out of business, a problem that seems to occur with increasing frequency.

Case Study

Company Leaves Town Without Warning

An inspector got an anonymous call on a Friday afternoon with the information that a large, local manufacturing company had just gone out of business. The management distributed paychecks that Friday, told all employees that the company was out of business, and told them not to bother showing up for work the following Monday.

The inspector arrived at the site that Friday afternoon to find the factory doors closed and the parking lot empty. The security guard was there and allowed the inspector access after the inspector presented his credentials and explained the purpose of the visit.

The owner had left the country with a large sum of money and was not expected back. The lower-level managers had no idea the closing was going to happen, so there was little information about the status of the closing.

The facility had been used for the manufacture of lawn mowers and snow blowers for over fifty years and there were numerous parts, partially and fully assembled lawn mowers and snow blowers, machinery for shaping and cutting metal, oils, paints, and offices, as can be seen in Figure 2.4.

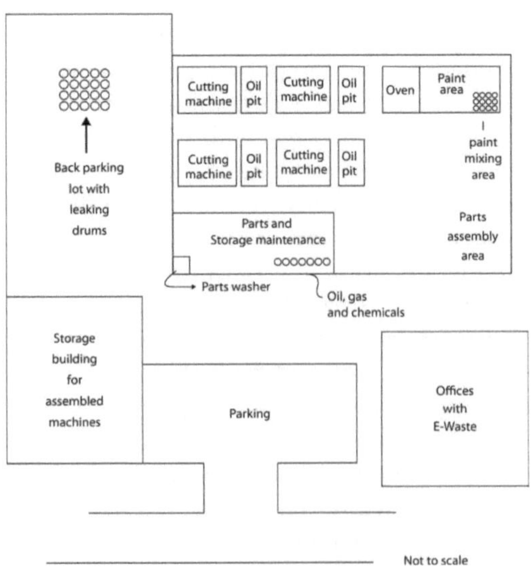

FIGURE 2.4 Company leaves town without warning.

Discussion:

Q: What kinds of hazardous wastes would be on site?

A: The manufacturing area would contain: paints, paint thinners, gasoline, inks, cutting oils, fluorescent lamps, acids, and bases. The offices would contain used electronics, mercury containing equipment (thermostats and switches), fluorescent lamps, and inks.

Q: If the owner has left the country and there is no person financially responsible or accountable, what can be done?

A: In this case, the outlook was bleak for environmental cleanup. There are no laws requiring businesses generating hazardous waste to have any financial surety or other mechanism if they decide to go out of business and abandon a site, provided the wastes are safely contained and there are no threats to the environment. Fortunately, in this case, the inspector discovered an oil spill from some open oil drums in the outdoor storage area of the facility. The inspector called in a petroleum spill, which alerted an emergency response. Under the authority of emergency response, the state was able to order an investigation, and the site was cleaned up by a company that had equipment stored on site.

There is a gap in the law in most states allowing hazardous waste generators to walk away from a facility without notifying the regulatory agency that the facility is being abandoned together with its contents, including hazardous wastes that had been properly stored and contained on site. If the wastes were stored and labeled properly, no hazardous waste storage violations would occur for up to ninety days. This is a long-standing challenge for waste regulators across the nation and can be solved only by legislation, which thus far has not been introduced.

In the next case study, we discover a businessman who simply refuses to acknowledge that he is generating hazardous waste. His insistence led to a host of problems for the business, the state regulatory agency, the local community, and, eventually, the federal government.

Case Study

Electroplater Who Claims He Has No Hazardous Waste

Many facilities have tried to be clever with the waste determinations, claiming there isn't really a waste because they still have use for it, or might have a use for it someday.

In one such case, an inspector visited an electroplating company (see Figure 2.5) where the owner was already under investigation for other unrelated behavior. The inspector obtained a search warrant based on employee complaints about poor waste management practices and potential health and safety issues.

Upon arrival, and after waiting for a uniformed officer to serve the search warrant, the inspector entered the facility and found two major issues:

1. Several hundred containers of electroplating chemicals were scattered throughout a very old and crumbling six story building. The owner insisted that none of the containers were waste. He claimed that they were still good plating solutions, and that they were being saved for later use. Upon testing the containers, it turned out that virtually all of the "good" plating solutions were actually spent to a point where they would be virtually useless in the future. The owner was accumulating these containers in order to avoid the cost of proper disposal.
2. The facility was still conducting electroplating, with numerous health and safety issues, including a "green fog" of fumes throughout the building from the plating operation. The workers complained of irritating fumes and respiratory problems. In addition, there was a large plating tank divided into two parts by one steel plate, a cyanide solution was contained on one side and an acid bath on the other. Any mixture of the two solutions would have caused a release of deadly cyanide gas.

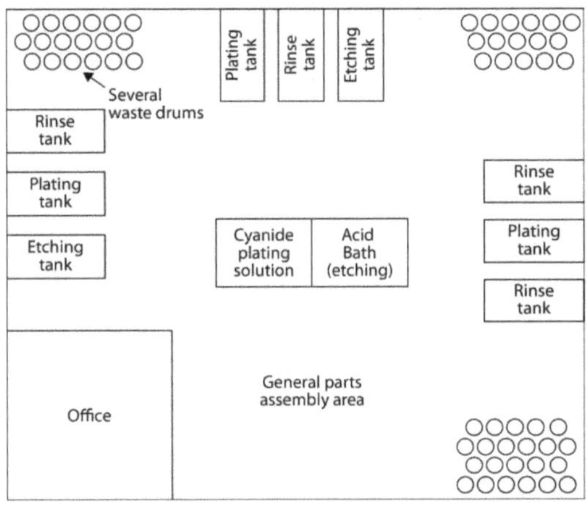

FIGURE 2.5 Electroplater who claimed to have no hazardous waste.

The inspector immediately contacted the Occupational Safety and Health Administration (OSHA), who immediately shut down the facility due to the imminent hazard associated with the tank containing acid and cyanide.

The owner claimed none of the containers contained waste. He stated that all of the containers contained valuable and useable plating solutions that he planned to use in the future. The sheer quantity of "product" was staggering, with hundreds of containers, some in very poor condition, scattered all over the six-story building.

To deal with the "product" that was actually waste, the federal government called the practice speculative accumulation and declared the entire property a hazardous waste cleanup site. The USEPA spent over $2 million securing the site and removing the containers. The building was demolished after the cleanup because it was deemed to be structurally unsound.

The previous case studies give a few examples of the issues associated with the identification of hazardous waste. The hazardous waste regulations can be interpreted in many different ways, and the safest determination is one agreed to by the USEPA and the home state regulatory agency staff. Any company that makes an independent waste identification or relies on a waste broker or vendor is taking a risk that may cause problems and possible additional costs.

In this chapter, you learned the primary responsibilities of generators and the categories of hazardous waste, discovered the relaxed standards for universal wastes, found out about mixed hazardous and radioactive wastes, and learned about special state hazardous wastes. You also learned why it's important to verify your waste determinations, and you read case studies depicting: that improper waste identification can be costly, a company that had left town without warning, and an electroplater who claimed no hazardous waste.

Update: The USEPA published a rule in December 2008 that gave colleges, universities, teaching hospitals, and non-profit research institutes flexibility to make hazardous waste determinations in order to more easily dispose of old and expired laboratory chemicals (adapted from USEPA website):

https://www.epa.gov/hwgenerators/managing-hazardous-waste-academic-laboratories-rulemaking

"On December 1, 2008, the USEPA added a new subpart, subpart K, to Title 40 of the Code of Federal Regulations (CFR) in Part 262, hazardous waste generator regulatory requirements. Subpart K is applicable to eligible academic entities, which are:

- *colleges and universities, and*
- *teaching hospitals and nonprofit research institutes that are either owned by or formally affiliated with a college or university.*

The purpose of the rule is to establish an optional, alternative set of regulations that allow eligible academic entities the flexibility to make hazardous waste determinations in the following settings:

- *in the laboratory;*
- *at an on-site central accumulation area; or*
- *at an on-site treatment, storage, or disposal facility (TSDF).*

This rule also provides incentives for eligible academic entities to clean-out old and expired chemicals that may pose unnecessary risk. Further, this rule requires eligible academic entities that choose to manage their hazardous waste according to the rule to develop a Laboratory Management Plan. The goal of this rule change was safer laboratory practices and increased awareness of hazardous waste management.

Eligible academic entities may also choose to opt out of the new standards in this rule and remain subject to the pre-existing hazardous waste generator requirements."

In the next chapter, you will learn more about U.S. hazardous waste management policy, see regulatory standards for categories of hazardous waste generators, hazardous waste transporters and treatment, storage and disposal facilities. You will also read case studies on sham recycling of waste tires, storage of incompatible materials, dangers of mixing incompatible wastes, universal waste confusion, a dry cleaner in New York City (local regulations supersede federal rules), and bulging drums in storage.

EXERCISES

1. Who is solely responsible for determining whether a hazardous waste is generated at a facility?
2. What are the four characteristics that make an unlisted waste hazardous?
3. What are the exact characteristics of
   a. An ignitable hazardous waste?
   b. A corrosive hazardous waste?
   c. A reactive hazardous waste?
   d. A toxic hazardous waste?
4. What is the F-list for hazardous waste?
5. What is the K-list for hazardous waste?
6. What is the P-list for hazardous waste?
7. What is the U-list for hazardous waste?
8. What is a universal waste?
9. Name four relaxed requirements for universal waste.
10. What are mixed wastes?
11. Is it possible for a waste stream to have more than one waste code?
    a. Give an example and explain.
12. Why should businesses always check with their home state about the identification of their wastes?
13. What specific subpart of the hazardous waste regulations was added in 2008? (Answer Subpart K)
14. What entities were affected by this new subpart? (Answer Academic entities like university laboratories and teaching hospitals)

REFERENCES

[EPA 11b] EPA Wastes Where You Live, online at *https://www.epa.gov/superfund/search-superfund-sites-where-you-live* (accessed April 2025)

[EPA 11c] EPA Contact Us Page, online at *https://www.epa.gov/aboutepa/forms/contact-epa* (accessed April 2025).

CHAPTER 3

Hazardous Waste Policy and Regulatory Requirements

In This Chapter

- U.S. hazardous waste management policy
- Regulatory standards for hazardous waste generators of different categories, hazardous waste transporters, and treatment, storage, and disposal facilities
- Case studies

3.1 HAZARDOUS WASTE MANAGEMENT POLICY

As discussed in Chapter 2, "Identification of Hazardous Waste," after evaluating the way hazardous waste was being managed in the United States subsequent to passing RCRA in 1976, Congress made policy decisions in the Pollution Prevention Act of 1990 (PPA) that directed the USEPA to steer industry away from disposing into landfills those wastes that could be otherwise treated. This was accomplished by setting treatment standards for all hazardous wastes at their points of generation.

Hazardous Waste Management Hierarchy

Although that policy was a step in the right direction, it wasn't until 1990 that Congress established a hazardous waste management hierarchy in the Pollution Prevention Act of 1990, stating under Section 6602 (b) that:

- Pollution should be prevented or reduced at the source whenever feasible.
- Pollution that cannot be prevented should be recycled in an environmentally safe manner whenever feasible.
- Pollution that cannot be prevented or recycled should be treated in an environmentally safe manner whenever feasible.
- Disposal or other release into the environment should be employed only as a last resort and should be conducted in an environmentally safe manner [PPA 90a].

The Hazardous Waste Management Hierarchy Broken Down

The Hazardous Waste Management Hierarchy meant that reduction of wastes at the source, waste minimization, waste exchange, and recycling must be considered and implemented by industries generating hazardous waste. These are advantageous requirements, partially because they benefit the environment, and also because these measures can result in significant savings to industries during manufacturing, thereby improving their profit margins.

The implementation of waste minimization requires a dedicated commitment by top-level management in any company to provide technical resources, financial backing, and buy-in at all employee levels. Making the commitment to put the environment ahead of other aspects of an organization is a critical step in making waste minimization effective. Waste minimization techniques sometimes overlap, but the basic categories are as follows:

Source reduction is defined by the Pollution Prevention Act as any practice that "(i) reduces the amount of any hazardous substance, pollutant, or contaminant entering any waste stream or otherwise released into the environment (including fugitive emissions) prior to recycling, treatment, or disposal; and (ii) reduces the hazards to public health and the environment associated with the release of such substances, pollutants, or contaminants. The term includes equipment or technology modifications, process or procedure modifications, reformulation or redesign of products, substitution of raw materials, and improvements in housekeeping, maintenance, training, or inventory control" [PPA 90b].

Other valuable methods of source reduction not mentioned in the Pollution Prevention Act are:

- Process quality control to make sure ingredients are refined and pure, using byproducts from other processes or wastes from other processes when possible, and making sure the process is environmentally friendly.
- Substitution of less toxic or more environmentally friendly materials wherever possible. Dilution of materials can be useful to produce a more dilute waste, as in electroplating or parts washing.
- Training of production staff and management can produce benefits in source reduction, particularly when there is complete buy-in and ownership of the entire manufacturing process, using employee recognition and incentive programs for suggesting and implementing changes that result in product improvement or waste reduction.

Pollution Prevention

According to the USEPA, "Pollution prevention means 'source reduction,' as defined under the Pollution Prevention Act, and other practices that reduce or eliminate the creation of pollutants through:

- increased efficiency in the use of raw materials, energy, water, or other resources, or
- protection of natural resources by conservation" [PPA 90a]. While this term is related to source reduction, this text places it slightly below source reduction because it implicitly indicates there is pollution to prevent, while source reduction means not creating pollution in the first instance.

Recycling/Reclamation/Reuse

Recycling is the reuse of waste materials for beneficial purposes. Recycled materials may be used, reused, or reclaimed. Reuse is returning waste materials (i.e., by-products or intermediates) to the original process as a substitute for a raw input material, or using those by-products or intermediates as input material to another process.

Reclamation involves running waste through a process to recover a useful or valuable material for future use. Examples of reclamation are: distilling a waste solvent to reclaim it, burning a waste oil to recover the heat value, and reclaiming acids and bases from plating solutions to use them for further plating or other manufacturing.

Recycling/reuse became heavily regulated by the state and federal governments, following some less-than-genuine, "sham" recycling operations conducted after the passage of RCRA in the late 1970s.

! TIP

It is wise to consult with the local state regulatory authority before commencing a recycling/reuse activity, just to make sure they agree that the proposed process meets their definition of recycling.

The following case study is an illustration of how much environmental damage can be caused by a sham recycler.

Case Study

Sham Recycling of Waste Tires

An inspector was directed to monitor compliance at a waste tire recycler (see Figure 3.1) who was shredding tires to be used for road base material and other purposes. The facility covered several acres, and the tires were stored in areas all over the site. Upon investigation, it was learned that the company had accepted tires at a price lower than its competitors, and when there were too many tires to shred, they buried whole tires in remote areas all over the site.

The site was declared a fire hazard because there were too many tires stored all over the site without proper rows for fire breaks or firefighting equipment. The site was cleaned up under the supervision of an investigation unit within the environmental agency, at a very large price. There have been so many waste tire sites throughout the nation, and the USEPA has developed a guidebook for cleanup of these sites [EPA 06].

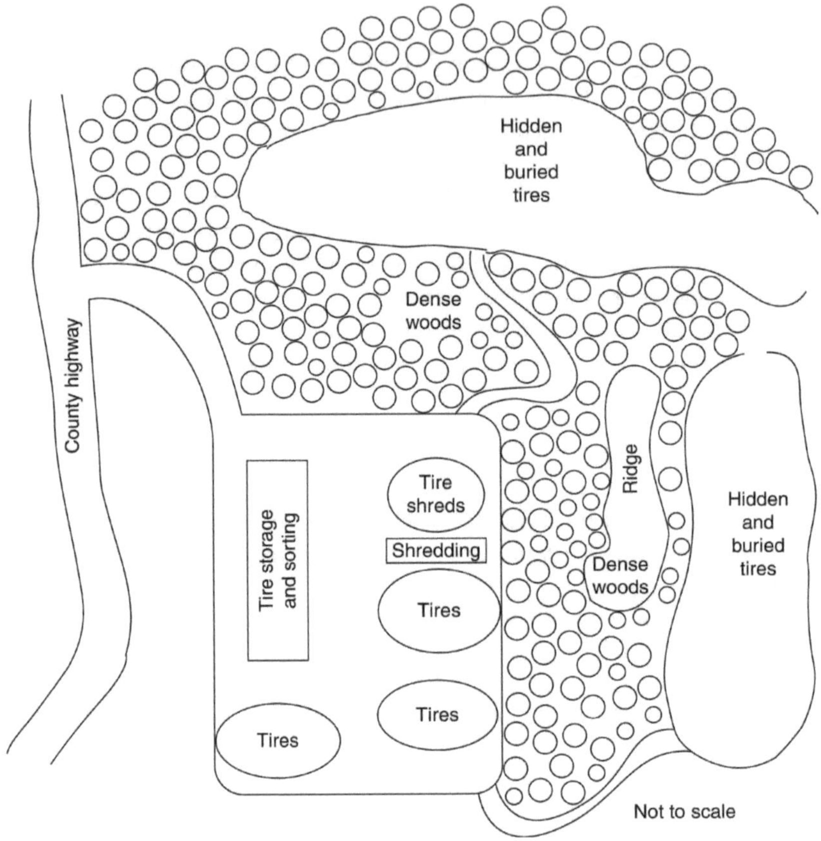

FIGURE 3.1 Sham tire recycler. (Drawing by author.)

Discussion: Although waste tires are not a hazardous waste, the burning of these tires creates a great deal of air and water pollution; as the tires melt, they release some of their base petroleum products. The petroleum products, once released from the tires, can seep into the ground, causing water pollution and potential hazardous waste sites.

3.2 REGULATORY STANDARDS FOR HAZARDOUS WASTE FACILITIES

Hazardous Waste Generators

After a facility determines its generator category, it must then follow the appropriate requirements for hazardous waste management, namely, storage, transportation, treatment, emergency response preparedness and record-keeping, which are outlined in this chapter for each generator category.

Update (Adapted from the USEPA regulatory summary at *https://www.epa.gov/hwgenerators/final-rule-hazardous-waste-generator-improvements*).

As discussed in the Foreword of this textbook, the USEPA made several (more than 60) changes to the Hazardous Waste Regulations in a rulemaking promulgated September 25, 2015. These changes have not gone into effect in states that have been granted authorization for the RCRA program. It may be a long time (up to several years in some states).

These regulatory revisions were intended to make the hazardous waste regulations easier to understand, like:

- changing the name Conditionally Exempt Small Quantity Generator (CESQG) to Very Small Quantity Generator (VSQG) to make it consistent with the other generator nomenclature (Very Small did not translate the meaning as well as VSQG);
- allowing VSQGs to send hazardous waste to a large quantity generator (LQG) that is under the control of the same person and consolidate it there before sending it on to management at a RCRA-designated facility, provided certain conditions are met;
- in some cases, organizations in industry, government, and academia are allowed to have satellite locations that qualify as VSQGs and that could take advantage of this provision to send their materials to an LQG within their company and ultimately to manage the hazardous waste in an environmentally sound manner rather than as an exempt waste.
- allowing a VSQG or a small quantity generator (SQG) to maintain its existing generator category in the case of an event in which the VSQG or SQG generates a quantity of hazardous waste in a calendar month that would otherwise bump the generator into a more stringent generator regulatory category (episodic waste generation). Under this provision, generators that satisfy the listed conditions do not have to comply with the more stringent generator standards when an unusual event such as a cleanout or an act of nature causes its generator category to temporarily increase;
- enhancing the safety of facilities that create hazardous waste and the response capabilities of emergency responders by improving risk communication;
- reorganize the regulations to make them easier to follow and make certain technical corrections.

Improvements of Environmental Protection

- Updating the emergency response and contingency planning provisions for SQGs and LQGs to include Local Emergency Planning Committees (LEPC) among those emergency planning organizations with which a generator may make response arrangements and to require that new and existing LQGs submit quick reference guides with the key information when they either develop or update their contingency plans to local responders for easy access during an event.
- Requiring periodic re-notification for SQGs every four years (SQGs only notify once under the current system).
- Revising the regulations for labeling and marking of containers and tanks to clearly indicate the hazards of the hazardous waste contained inside.

Provisions to improve generator compliance (Housekeeping)

- Clarifying inconsistent guidance on which generator category applies when a generator generates both acute and non-acute hazardous waste in a calendar month.
- Revising the regulations for completing the RCRA biennial report to be consistent with the current instructions distributed with the form.

Reorganization of the hazardous waste generator regulations

- Moving the VSQG regulations from section 261.5 of Title 40 of the Code of Federal Regulations (CFR) into 40 CFR part 262, where the regulations for SQGs and LQGs are located.
- Moving a number of the generator regulations that are currently located in other parts of the hazardous waste standards into 40 CFR part 262 to replace the current lists of cross references.

Technical corrections

- Correcting inadvertent errors in the regulations, obsolete programs, and unclear citations."

e-Manifest Rule

It is important for the reader to know that the changes to the Hazardous Waste Rules above will go into effect in each state after the laws and regulations become effective in that particular state. This is not true for the new e-Manifest Rule, which becomes effective nationwide on June 30, 2018. As discussed in the Introduction, these rules have been in development at the USEPA since 2012, and they will be rolled out nationwide at one point in time so the USEPA can more easily track hazardous waste shipments anywhere in the United States and its provinces, and international shipments to and from the U.S.

A complete description of the e-manifest system can be found at:

https://www.epa.gov/e-manifest/learn-about-hazardous-waste-electronic-manifest-system-e-manifest

The fees for the new system and the process for verifying signatures are at:

https://www.epa.gov/e-manifest/final-rule-user-fees-electronic-hazardous-waste-manifest-system-e-manifest-and-amendments

Categories of Hazardous Waste Generators

Nonregulated Handler (NRH)

While the USEPA doesn't list this as a regulated generator category, the NRH represents an important regulatory status used by regulators and compliance inspectors.

All companies and businesses, whether or not they generate hazardous waste, must make and document their hazardous waste determinations. These determinations should be verified with the home state regulatory authority, and documented when possible. These records should be kept so that they can be produced in the event of a compliance inspection.

NOTE

In some cases, a facility can have hazardous wastes generated on site and still be considered a nonregulated handler. For example, if a nonregulated handler has fluorescent lamps or other electronic equipment in their office(s), and the used lamps and other electronic equipment are changed out and managed by a landlord or property management company, it is possible that the home state may not require the business to count the waste as generated by them. In this case, the landlord is the generator of the hazardous waste; the company is not. The landlord enjoys the relaxed regulatory standards if the universal wastes are managed properly.

Very Small Quantity Generators (VSQG)

A VSQG generates no more than 100 kilograms (kg) or 220 pounds (lbs.) per month of hazardous waste, or 1 kg (2.2 lbs.) of acute hazardous waste per month (P-List).

The VSQG regulatory requirements are summarized from the EPA checklist below [EPA 11d]:

- The VSQG must make and document all hazardous waste determinations. There were frequent errors made in waste determinations by many VSQGs in the experience of the author. Although not required, it is easy to avoid these errors by verifying the determination with the home state regulatory authority.
- The VSQG must not accumulate more than 1,000 KG (2,200 LB) non-acute hazardous waste on site and no more than 1 KG (2.2 LB) acute hazardous waste on site. If these storage limits are exceeded and observed by an inspector, the facility will be inspected as a Small Quantity Generator, facing many more numerous and rigorous regulatory requirements.
- The VSQG must treat and dispose of their hazardous waste in an on-site facility or ensure delivery to an authorized off-site facility. This requires some careful research on the part of the VSQG to verify that the facility is authorized to take their waste. They should never take the word of any waste broker or handler about this, but should demand to see the written permits.
- The VSQG may either self-transport its waste to an authorized facility or ship its waste to an authorized facility using a licensed transporter, requiring all transporters and receiving facilities to provide written proof of permits, or checking with the home state regulatory authority for permits.

NOTE

In determining hazardous waste generator categories, universal wastes generated at the facility do not have to be counted as part of the hazardous waste generated, provided the waste is managed in conformance with the universal waste rules. If the universal waste is not managed properly (i.e., improper storage, breakage, or other ineffective management), it is counted as hazardous waste and is added to the total hazardous waste generation for the facility and sometimes can increase the regulatory requirements (and fees). Other wastes may not count toward hazardous waste generation either, such as neutralized wastes, etc.

Small Quantity Generators (SQG)

A SQG generates more than 100 kg (220 lbs.) and no more than 1,000 kg (2,200 lbs.) of non-acute hazardous waste per month, or no more than 1 kg (2.2 lbs.) of acute hazardous wastes per month.

The SQG regulatory requirements are summarized from the EPA checklist as follows [EPA 11e].

- SQGs must make and document hazardous waste determinations. Again, this was the source of frequent errors in the experience of the author. Although not required, it is easy to avoid these errors by verifying the determination with their home state regulatory authority. No facility should rely on a waste broker or handler to make the waste determination for them, because the facility is solely responsible for this.
- When they first learn that they generate hazardous wastes, SQGs must obtain an USEPA Generator ID number by contacting the appropriate USEPA regional office (found at *http://www.epa.gov/aboutepa/where.html*). The USEPA ID number will remain with the facility for the life of the facility. Note that if the facility has more than one site on different, non-contiguous parcels of land, a separate USEPA ID number is required for each site.
- SQGs must arrange for the shipment of their hazardous waste using a licensed hazardous waste transporter. They must verify the transporter is licensed and keep a copy of the transporter permit(s) on file, in case they are inspected. The SQG should not ship the waste with a transporter unless the driver has a copy of the transporter permit in his truck, and the generator has a copy of the permit in its records.
- SQGs can accumulate no more than 6,000 kg of non-acute hazardous waste or no more than 1 kg of acute hazardous waste on site. If they do, they are in violation of their storage requirements, and are then regulated as a Large Quantity Generator (LQG), a category that has to follow much more stringent regulatory requirements.
- SQGs must have their hazardous wastes shipped off site within 180 days of the date of generation, unless the ultimate disposal facility is no further than 200 miles away. The SQG may store the waste up to 270 days if the disposal facility is further than 200 miles away. This must be documented for possible inspections. Exceeding these time frames also causes the SQG to be regulated as an unpermitted facility, subjecting the facility to possible enforcement actions, and penalties.
- The SQG must meet the requirements for preparedness and prevention, and emergency measures listed below, making sure one employee is always on site or on call to coordinate emergency measures and to post the following:
  - The SQG must post the name and phone number of the facility's emergency coordinator next to the telephone.
  - The SQG must post the location of all fire extinguishers and spill control equipment next to telephone.
  - The SQG must post the telephone number of fire department next to the telephone unless the facility has a direct alarm.

NOTE

The previous three requirements, marked with a O, are unique to SQGs, meaning no other type or class of facility is required to meet them.

- The SQG must conduct training of staff on waste handling or emergency measures, and, for future inspections, it is highly recommended the facility keep records of the content of that training, when it was held, and who received the training.
- The SQG must conform with all accumulation and storage requirements listed below:
  - All containers must be in good condition and not leaking (this includes the requirements that the containers be stored in a safe location).
  - All wastes must be stored in containers made of compatible materials (i.e., corrosive wastes must be stored in polyethylene, glass, or some other corrosion resistant material).
  - All containers that are not in use are closed (the cover on or bung installed). "In use" has been the topic of many enforcement discussions, but generally is enforced if the manufacturing process that generates the waste is not operating when the inspector is there.
  - All containers must be opened, handled, and stored to prevent leaks. This is a broad requirement for container management, and includes keeping stored containers safe from tipping over, being hit by equipment, or keeping aqueous wastes from freezing;
  - Each container must be marked with the words "Hazardous Waste" and with other words to identify the contents.

! TIP

This was a common error at the author's inspections, in that the facility would often write the words "Hazardous Waste" or words to describe the wastes, but not both. The requirement is for both. It is very helpful to the inspector for the facility to mark down the EPA waste codes as well.

  - Waste containers for each separate waste must be dated, and can be accumulated for no longer than three days in accumulation areas, after they are full. This rule can be confusing, in that it is acceptable to have more than one or even several 55-gallon accumulation containers of different waste types in the same areas.

NOTE

For each waste stream, the three-day clock to move the waste out of the accumulation area starts when the container is full. The SQG is required to place on the container the date when each 55-gallon container was filled, and then has three full days to move the container to the fewer than 90/180/270-day storage area. The inspector checks the full date on containers in accumulation areas to make sure that they have been there for no longer than three days.

For areas with fewer than 180 days of storage:

- The SQG must clearly mark the date that the container was placed in the "less than 180 day storage area" – on each container. This is important because it allows the inspector to determine if the SQG is within the 180-day storage window. Without this information, the inspector can call this a violation of the storage time limit.

- The SQG must mark each container with the words "Hazardous Waste" and with other words to identify the contents. Again, the requirement is to mark each container with both sets of words, or it will be a violation.
- All containers must be in good condition and not in danger of leaking.

! TIP

It is a small investment to purchase new or refurbished containers in order to make sure the containers do not leak from corrosion or weakness at the seams of the containers. It is also unwise to employ containers previously used for other wastes or materials unless they are thoroughly emptied and cleaned, inside and out. If employing containers previously used to store different wastes or materials, the facility should remove or paint over all previous markings to avoid confusion and possible labeling/marking violations

---

- The hazardous waste must be stored in containers made of compatible materials.

! TIP

Also, it is a good idea to have separate storage areas for incompatible wastes, with separate containment areas should they both leak, for health and safety purposes.

- All containers must be closed unless in use (being emptied or filled).
- The SQG must ensure that all containers are opened, handled, and stored to prevent leaks.
- The container storage area must be inspected at least once per week. These inspections should be recorded on some type of log sheet to prove the inspections were made. The log sheet should include the requirements listed above. While the regulations do not require that SQGs keep facility records, it is a good business practice to keep these records.
- If they use a tank or tanks for hazardous waste storage, the SQG must meet all tank storage requirements listed in EPA regulations, Parts 264 and 265. (These are listed in Appendix 2 on the companion files.)
- The SQG must ensure that all waste shipments are accompanied by a hazardous waste manifest, as in Figure 3.2 [EPA 11e].

Note that this form will be transitioned to an electronic manifest on June 30, 2018, as mandated by the USEPA e-manifest regulations. More detailed information on the implementation of this new program is forthcoming from the USEPA.

It is very important that the generator be aware and support all of the information entered on the hazardous waste manifest because it is the key document to prove their waste arrived at its final destination.

NOTE

These manifests contain a great deal of information about the generator, the waste, how it is to be managed, and its destination. An improperly completed manifest can create serious violations, especially if the hauler is stopped for a routine waste check or is involved in an accident. As discussed above, in the waste determination section, the generator is solely responsible for completing the manifest and ensuring its accuracy. In the eyes of the regulator, waste broker or handlers are not responsible for correctly completing the manifest form.

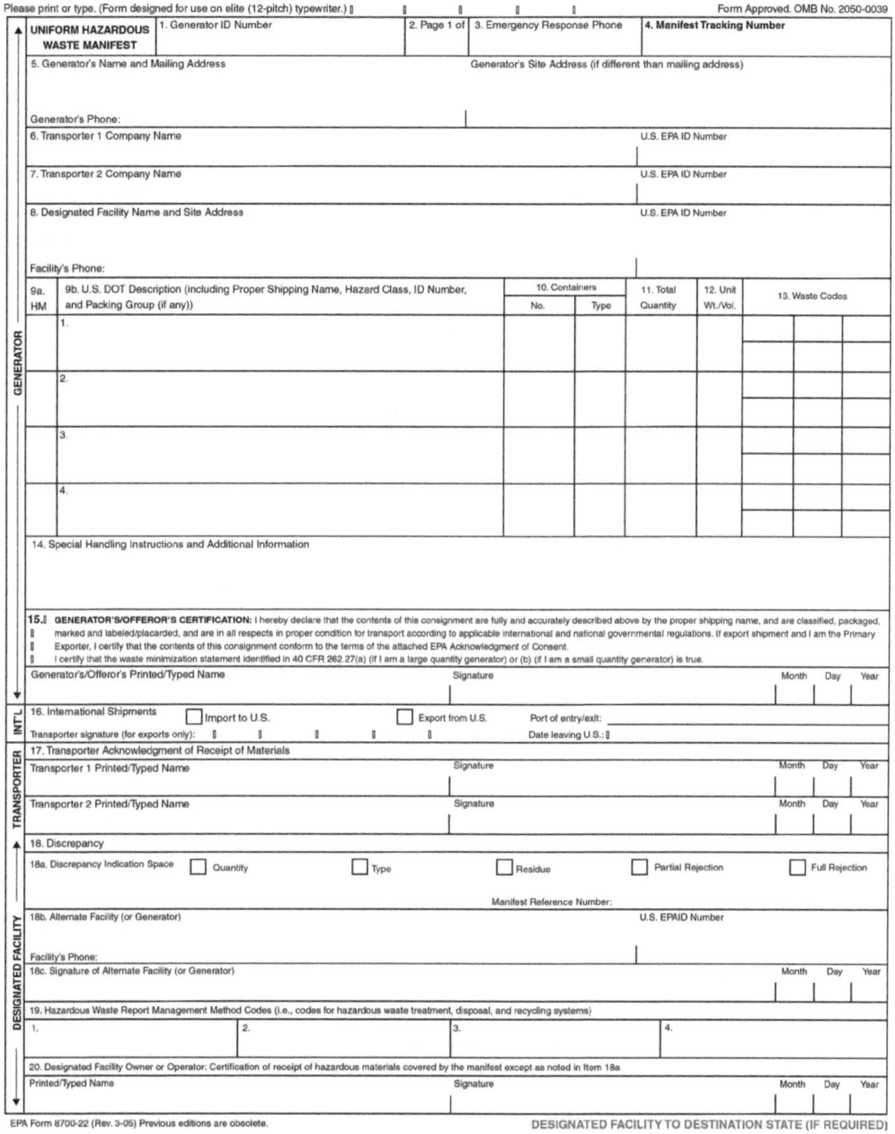
Please print or type. (Form designed for use on elite (12-pitch) typewriter.)
Form Approved. OMB No. 2050-0039

UNIFORM HAZARDOUS WASTE MANIFEST | 1. Generator ID Number | 2. Page 1 of | 3. Emergency Response Phone | 4. Manifest Tracking Number

5. Generator's Name and Mailing Address | Generator's Site Address (if different than mailing address)
Generator's Phone:

6. Transporter 1 Company Name | U.S. EPA ID Number

7. Transporter 2 Company Name | U.S. EPA ID Number

8. Designated Facility Name and Site Address | U.S. EPA ID Number
Facility's Phone:

9a. HM	9b. U.S. DOT Description (including Proper Shipping Name, Hazard Class, ID Number, and Packing Group (if any))	10. Containers No.	10. Containers Type	11. Total Quantity	12. Unit Wt./Vol.	13. Waste Codes
	1.					
	2.					
	3.					
	4.					

14. Special Handling Instructions and Additional Information

15. GENERATOR'S/OFFEROR'S CERTIFICATION: I hereby declare that the contents of this consignment are fully and accurately described above by the proper shipping name, and are classified, packaged, marked and labeled/placarded, and are in all respects in proper condition for transport according to applicable international and national governmental regulations. If export shipment and I am the Primary Exporter, I certify that the contents of this consignment conform to the terms of the attached EPA Acknowledgment of Consent.
I certify that the waste minimization statement identified in 40 CFR 262.27(a) (if I am a large quantity generator) or (b) (if I am a small quantity generator) is true.
Generator's/Offeror's Printed/Typed Name | Signature | Month Day Year

GENERATOR

INT'L
16. International Shipments ☐ Import to U.S. ☐ Export from U.S. Port of entry/exit:
Transporter signature (for exports only): Date leaving U.S.:

TRANSPORTER
17. Transporter Acknowledgment of Receipt of Materials
Transporter 1 Printed/Typed Name | Signature | Month Day Year
Transporter 2 Printed/Typed Name | Signature | Month Day Year

DESIGNATED FACILITY
18. Discrepancy
18a. Discrepancy Indication Space ☐ Quantity ☐ Type ☐ Residue ☐ Partial Rejection ☐ Full Rejection
Manifest Reference Number:
18b. Alternate Facility (or Generator) | U.S. EPA ID Number
Facility's Phone:
18c. Signature of Alternate Facility (or Generator) | Month Day Year
19. Hazardous Waste Report Management Method Codes (i.e., codes for hazardous waste treatment, disposal, and recycling systems)
1. | 2. | 3. | 4.
20. Designated Facility Owner or Operator: Certification of receipt of hazardous materials covered by the manifest except as noted in Item 18a
Printed/Typed Name | Signature | Month Day Year

EPA Form 8700-22 (Rev. 3-05) Previous editions are obsolete.
DESIGNATED FACILITY TO DESTINATION STATE (IF REQUIRED)

FIGURE 3.2 Hazardous waste manifest form. (From EPA Web site at: *http://www.epa.gov/osw/hazard/transportation/manifest/pdf/newform.pdf.*)

In addition:

- The SQG must make sure each hazardous waste manifest is accompanied by a properly completed land disposal restriction form. The SQG must maintain copies of all manifests and exception reports for three years, keep records of test results and determinations for three years, keep written communication with treatment storage and disposal facility, keep written proof that the transporter(s) are authorized to deliver their waste, and must meet preparedness and prevention requirements [EPA 11d].

NOTE *Universal wastes generated at the facility do not have to be counted as part of the hazardous waste generated, provided the waste is managed in conformance with the universal waste rules.*

NOTE *Hazardous wastes that are identified as hazardous because of corrosivity (pH ≤ 2 or ≥12.5) or liquids capable of corroding containers may be in many cases neutralized on-site and not be counted in the categorization of the generator. This can be significant and in some cases, change the category of a generator to a lower level (i.e., LQG to SQG or SQG to VSQG), reducing regulatory requirements and costs significantly. See detailed discussion after theTSDF summary below.*

Large Quantity Generators (LQG)

LQGs generate ≥ 1,000 kg (2,200 lbs.) of hazardous waste per month, or greater than 1 kg (2.2 lbs.) acute hazardous waste per month, and stores more than 1,000 kg of hazardous waste and more than 1 kg of acute hazardous waste.

NOTE *Most of the requirements for LQGs are the same as the SQG requirements, with the some exceptions (such as of the three items related to the posting of different information by the telephone listed previously under SQG requirements).*

An abbreviated list of the additional requirements for LQGs is provided by the EPA checklist [EPA 11d] as follows:

A LQG has only 90 days to ship a waste off site, as opposed to 180/270 days for a SQG.

- LQGs must fill out a Generator Biennial Report (or Annual, depending on the home state) and pay the appropriate generator fees and taxes (depending on the home state regulations).
- LQGs must conduct personnel training of all staff in hazardous waste management at the facility, and maintain proof of this training for three years. These records must include the job title for each position, written job description, names of the people filling the jobs, a written description of the training, including contingency-plan training.
- The LQG must meet all regulatory requirements for preparedness and prevention, including internal alarms, communication devices, fire extinguishers, adequate aisle space to allow emergency access, and provide proof that emergency agreements/arrangements are in place with police departments, fire departments, and local hospitals.
- The LQG must have a detailed written contingency plan that includes what facility personnel must do in emergencies, arrangements with emergency authorities, names and addresses and home phone numbers of facility emergency coordinators, a list of all emergency equipment at the facility, and an evacuation plan. This plan must be distributed to all local police, fire departments, and hospitals, and updated as necessary.

As with SQGs, universal wastes generated at the facility do not have to be counted as part of the hazardous waste generated, provided the waste is managed in conformance with the universal waste rules.

Hazardous Waste Transporters

Transporters are allowed to store and transfer hazardous wastes from one vehicle to another for up to 10 days, provided they: obtain and maintain permits to transport hazardous waste, obtain a USEPA ID number, comply with all licensed transporter permit requirements, store waste on site for up to 10 days, and record all transfers from one vehicle to another on hazardous waste manifests for all shipments.

NOTE *Wastes generated by spills or containment and cleanup of spills are normally considered generated by the transporter facility. Exceptions may occur, where the container(s) used by the generating facility were damaged or leaking prior to pick up [EPA 11d].*

Interim Status Treatment, Storage, and Disposal Facilities (TSDFs)

Interim Status TSDFs are facilities that were in existence when the hazardous waste regulations went into effect or who subsequently became subject to hazardous waste permitting due to changes in the regulations. There are very few of these facilities nationwide. However, those that do exist are required to submit a permit application, and until they do, the existing facility is allowed to continue operation until a final permit is obtained. The requirements for an interim status facility are similar to a LQG, with added requirements for notifications when they receive waste, waste analysis plans, heightened security requirements, groundwater monitoring, closure and post closure plans, and financial sureties in the event that they close and unit (e.g., tank, container, containment building, etc.) specific requirements.

Treatment, Storage, and Disposal Facilities (TSDFs)

TSDFs operate under a hazardous waste permit that is issued by the regulating state or the federal government, following a rigorous review and a public review process.

Basic requirements for all TSDFs are: they must obtain, maintain, and comply with their hazardous waste permit; if hazardous waste is generated, the TSDF must comply with all generator requirements; if the TSDF is a transporter of wastes, it must comply with all transporter requirements; the TSDF must pay TSDF fees, generator fees (if necessary), and keep records; and the TSDF must train staff and maintain training records.

Exemptions from Permitting

There are certain activities that may exempt facilities (or portions of a facilities) from needing hazardous waste permits, as listed below:

- Cleanup at contaminated hazardous waste sites. The cleanup contractor is required to meet the applicable or relevant and appropriate requirements (ARARs) in RCRA.
- Facilities do not store hazardous waste generated on site in containers or tanks for longer than 90 days.
- SQGs do not store more than 1,000 kg – 6,000 kg (2200 – 13,200 lb.) on site at any time.
- Facilities storing and recycling recyclable materials listed in the regulations, such as certain waste fuels and waste oils.

Using Waste Neutralization of Corrosive Wastes to Reduce Generation (and Possibly Category)

Hazardous waste generators have the option to consider the neutralization of some hazardous wastes to reduce both the amount of hazardous generated at the facility and disposal costs. The hazardous waste regulations allow a generator to reclassify a hazardous waste that was classified as a hazardous waste due to its characteristics (ignitable, corrosive, reactive, or toxic) to a non-hazardous waste if the waste no longer exhibits any of the characteristics of a hazardous waste. This reclassification does not apply to any listed hazardous waste.

The facility can reclassify the corrosive waste by changing the characteristics of the waste through neutralization. To be a non-hazardous waste, the corrosive waste must have a pH greater than 2.0 and less than 12.5. In addition, the waste can no longer meet any of the other characteristics of a hazardous waste (ignitable, corrosive, reactive, or toxic).

The hazardous waste regulations contain three potential exemptions concerning neutralization.

1. Totally enclosed treatment facilities

   The regulations define a totally enclosed treatment facility as a facility that is directly connected to an industrial process and which is constructed and operated in a manner, which prevents the release of any hazardous waste or any constituent to the environment during treatment. An example given in the regulation is a pipe in which waste acid is neutralized. Other examples would be an air scrubber that is design to neutralize the scrubber water or an inline tank where the process stream is neutralized.

   The regulations view this as a part of the process, and there are no regulatory requirements until the material leaves the process and is declared a waste. At this point, the waste material is characterized as a hazardous or non-hazardous waste.

2. The on-site treatment of hazardous waste in the same tanks or containers that are used for the accumulation and storage of such waste.

   Under this exemption, the waste can be neutralized in the container or tank used for accumulation or storage.

   Under this exemption, the regulations require:

   - The date on which each period of accumulation begins is clearly marked and visible for inspection on each container;
   - A label or sign stating "Hazardous Waste " must identify all area, tanks and containers used to accumulate hazardous waste. In addition, tanks and containers must be marked with other words to identify their contents. This label can be removed once the material has been neutralized and determined not to be a hazardous waste.
   - Each Container is properly labeled and marked for shipment, if sent off site.
   - The personnel training requirements are satisfied.
   - Emergency preparedness procedures are to be in place.
   - Written Contingency plans are in-place.
   - The facility must submit a written notification to the regional office stating that it qualifies for this exemption. Any tank and container storage areas are to be located within a secondary containment system.

3. Elementary neutralization unit (6 NYCRR 373-1.1(d)(xii))

The regulations define an elementary neutralization unit as a tank, tank system or container that is used to neutralize a waste that is hazardous because the waste only exhibits the characteristic of corrosivity.

Under this exemption, the elementary neutralization unit is designed to neutralize the waste that has been accumulated. The regulations require:

- The date on which each period of accumulation begins is clearly marked and visible for inspection on each container;
- A label or sign stating "Hazardous Waste "must identify all area, tanks and containers used to neutralized the hazardous waste. In addition, tanks and containers must be marked with other words to identify their contents.
- The personnel training requirements are satisfied.
- Emergency preparedness procedures are to be in place are in place.
- Written Contingency plans are in-place.
- The containers and tanks are managed in accordance with the regulations.
- Waste Analysis Plan.
- The tank and container storage areas are to be located within a secondary containment system.
- A closure plan must be prepared.

When the neutralized waste no longer meets any of the characteristics of a hazardous waste, it can then be reclassified as a non-hazardous waste and sent off-site.

Any wastewater that is generated by a facility can be discharged to a municipal sewer system following approval from the receiving treatment facility, or through a wastewater discharge permit. To obtain approval, an application is made to the facility describing the wastewater characteristics, quantity and frequency that the wastewater would be discharged to their facility. Upon review, the wastewater treatment facility would determine if the facility can receive the wastewater and if acceptable, issue a permit. The permit would contain what could be discharge, reporting requirements, and any other applicable conditions.

3.3 CASE STUDIES

Case Study

Company changes generator status and saves big using waste neutralization.

A company regulated as a Large Quantity generator was generating thousands of gallons of corrosive aqueous waste ($pH \leq 2$), placing it in 55 gallon drums, and manifesting it off-site as hazardous waste.

This was an expensive method to manage the waste, not only from an economic perspective, but also from a risk basis, since highly corrosive liquids were being transported on public highways.

The company was advised that the corrosive aqueous wastes could be legally neutralized on-site, reducing costs of transportation and disposal. The company installed waste neutralization equipment and was delighted to find that the neutralized waste no longer counted toward their monthly hazardous waste generation. This allowed the company to change its regulatory status from Large Quantity Generator to Small Quantity Generator, saving them:

- a great deal of money in waste generation regulatory fees;
- more money in reduced transportation and treatment costs;
- significantly reduced regulatory requirements (LQG to SQG); and
- reduced risks and liability associated with transportation and off-site disposal.

Some companies have taken advantage of this regulatory break, but many smaller companies have not.

Case Study

Storage of Incompatible Hazardous Wastes in Improper Containers

One of the most dangerous mistakes made in the storage of hazardous waste is storing acids and bases in the same containment areas. Figure 3.3 shows one facility where a hazardous waste inspector found six 55-gallon steel drums of D002 acid waste next to six 55-gallon steel drums of D002 basic (caustic) waste and six 55-gallon drums of solvents, all stored in the same secondary containment area.

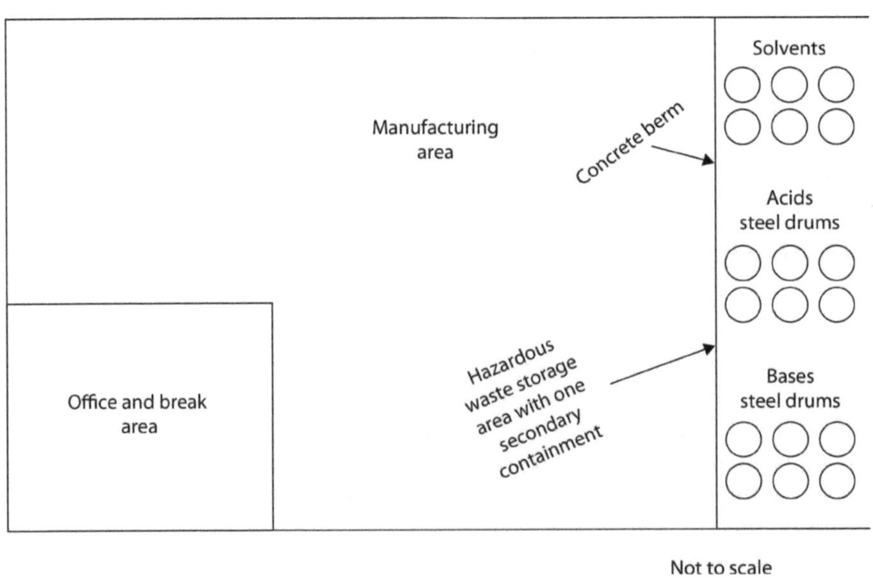

FIGURE 3.3 Incompatible wastes stored in same secondary containment area. (Drawing by author.)

The first violation was the storage of wastes using containers that were not compatible with the wastes. The acids and bases are both corrosive, and should never be stored in steel containers.

The second violation was keeping the three incompatible wastes (acids, bases, and solvents) in the same secondary containment area. If the drums had leaked or ruptured, any of the three types of chemicals could have reacted violently, creating possible fumes, explosions, and other hazardous situations.

The company was cited for improper storage of incompatible wastes, and was instructed to transfer the acids and bases into containers made of compatible materials, and to place each of the different types of wastes in separate secondary containment areas.

Discussion: Hazardous waste inspectors find this problem at many site visits, and the issue extends to improper storage of incompatible materials. In one of the case studies in Chapter 2, Electroplater Who Claims He Has No Hazardous Waste, the facility had a production tank with acids and cyanide solutions in the same tank, separated only by a thin sheet of steel. Storage issues such as these are highly dangerous, and owners/operators of production facilities and waste storage areas must be vigilant in order to avoid possible serious incidents.

Below is an example of what can happen if incompatible wastes or materials are mixed together.

Case Study

Dangers of Mixing Incompatible Wastes

Hazardous waste activities can be very dangerous, especially when something goes wrong. In the activities associated with consolidating wastes, facilities and waste managers must be extremely careful when mixing different contents of more than one container into a single container. The mixing of incompatible wastes has resulted in many incidents, and an inspector was involved in the aftermath of one such incident (see Figure 3.4). In April 2002, workers were consolidating some small (15-gallon) unlabeled waste containers into larger (55-gallon) containers at a sign shop in Manhattan, NY, when they accidentally pumped a few gallons of nitric acid into a 55-gallon drum containing lacquer thinner (organics). The drum started hissing, so the employees ran out of the basement just before the drum exploded. Thankfully, no one was killed, but 36 people were injured, including 14 members of the public and six firefighters. The force of the explosion caused a partial collapse of the building, knocking out most of the windows in the six-story building as the concussion went up the elevator shaft and literally blew the windows out to the streets below.

FIGURE 3.4 Aftermath of Kaltech Explosion. (From U.S. Chemical Safety and Hazard Investigation Board, April 2002, online at *http://www.csb.gov/assets/news/image/Kaltech1_0001.jpg*.)

The sign shop was in the basement of the building, and the rest of the building was used as offices by service firms, professionals, and other businesses. Many of the other occupants of the building claimed they didn't know hazardous wastes were managed in the building. A summary of the investigation log with the recommendations of the Chemical Safety Board is on the Web [CSB 02].

Following this incident, a group of hazardous waste inspectors from throughout New York State were directed to visit several hazardous waste facilities in Manhattan to make sure they were in compliance with all of the hazardous waste rules and regulations, in order to prevent possible future incidents. The focus was on "combined use" buildings, where multiple businesses or residences occupied the same building where hazardous wastes were generated.

There has been a serious lack of communication from government to small businesses that generate hazardous waste. From the time RCRA was passed (1976) until now, many small businesses have not been aware of the requirements contained in the hazardous waste rules and regulations.

Following is a case study that depicts the lack of communication and resulting confusion at businesses concerning the existence of universal wastes and the advantages of declaring certain wastes as universal wastes.

Case Study

Universal Waste Confusion

When inspected, many businesses seem surprised to learn that they are subject to hazardous waste rules and regulations, particularly for the generation of fluorescent lamps waste. The lamps were a hazardous waste when RCRA was first passed in 1976, because each lamp contained enough mercury to make them fail the Toxicity Characteristic Leaching Procedure (TCLP). Any facility who replaced fluorescent lamps became a generator of D009 hazardous waste (see Chapter 1). The Universal Waste rules for lamps were not written until 1999, so these used lamps were regulated as hazardous waste.

Unless they had been inspected previously, many of the small businesses were not aware that it was illegal to dispose of the lamps with their municipal trash. Each inspector was responsible for warning the businesses and advising them to manage their lamps as hazardous waste. The situation improved markedly when the Universal Waste Rule for lamps was passed, but many businesses continued to be uninformed of lamp disposal requirements until they were inspected. After the Universal Waste rule was put in place in 1999, the recycling of lamps increased, but many industries did not take advantage of these relaxed standards.

The pollution caused by the improper disposal of fluorescent lamps was hard to quantify, but there was one source that produced a very large quantity of lamps—the tanning booth business. More than one inspector responded to complaints about tanning businesses and their disposal of lamps. On one occasion, the inspector found over 1,000 bulbs in a municipal waste dumpster, waiting for pickup. The inspector called a police officer, who wrote the owner of the tanning business a ticket. The inspector advised the owner to recycle the bulbs that had been put into the dumpster and to recycle all bulbs in the future. The waste hauler was also advised by the officer to refuse to pick up lamps in the future.

Part of this problem has been handled with public outreach efforts and the development of lamps with lower levels of mercury, but the disposal of mercury in the municipal waste stream continues to be a significant problem.

Although an individual state has to have hazardous waste rules and regulations that are at least as stringent as the federal government, sometimes local governments have even more stringent rules and regulations than their home states. This can be very confusing for the industries within that municipality, and they can pay a price for those differences. This case study concerns one of those situations.

Case Study

Dry Cleaner in New York City

(Local regulations supersede federal rules.)

Beyond the issue of individual states sometimes having more stringent hazardous waste regulations than the USEPA, some local governmental districts within the individual states have regulations that are even more stringent than those of the home state.

During a hazardous waste inspection of a dry cleaning establishment in New York City (Manhattan), a hazardous waste inspector conducted an inspection and found the facility to be in substantial compliance with the federal and state rules and regulations.

The federal and state regulations allow the discharge of the waste from dry cleaner's separators to a public sewer. However, in Manhattan, it was illegal for dry cleaners to discharge this waste into a public sewer, so the dry cleaner had to go through an additional rigorous effort to prove this waste was being collected, accumulated, and stored for off-site disposal. A separate pretreatment permit was required for discharge into the publicly owned sewer system.

Discussion: In general, as discussed in Chapter 2, hazardous wastes are normally regulated under RCRA to the point where they are either stored or treated. RCRA regulations do not cover discharge to a public sewer. Other municipalities have differing rules about hazardous waste management, so it is always necessary for generators of hazardous waste to check with their local municipal authorities to know if special local rules exist.

The next case study discusses hazardous waste storage precautions industry should take in those climates where freezing temperatures are a possibility.

Case Study

Bulging Drums in Storage

On a very cold day in the Northeastern section of New York State, an inspector visited a permitted storage facility when the temperature was well below freezing (about 10 °F).

The facility was in compliance with most of the state and federal regulations and their hazardous waste permit conditions. The only exception was a number of 55-gallon steel drums that were stored outside, bulging at the bottoms and tops. It turned out the wastes contained aqueous wastes and the water in the drums had frozen, causing them to bulge.

The facility paid a monetary penalty and corrected the situation by placing all future aqueous waste in storage buildings that were maintained temperatures above 32 °F all the time.

Discussion: Similar situations occur throughout the areas where the temperature goes well below 32 °F, so facility managers are well advised to keep their aqueous wastes in heated storage. Drums may also bulge from high internal pressure, caused by very high temperatures or chemical reactions, as in Figure 3.5, which is from a report about the 2002 Kaltech investigation mentioned earlier.

FIGURE 3.5 Bulging steel drum. (From Investigation Report No. 2002-02-I-NY, Figure 4, Page 20, U.S. Chemical Safety and Hazard Investigation Board, September 2003.)

This bulging drum issue has caused numerous violations in areas of the United States where the temperatures can stay below freezing for extended periods of time or where wastes might be exposed to high pressure. An inspector is not going to cite a violation for improper storage unless the container is affected.

Water and any other compound that expands when it freezes (i.e., acetic acid, elemental silicone, etc.) are wastes of concern for causing bulging containers, because their crystallization upon freezing causes volume expansion.

In the next case study, we will look into a very serious aisle-space violation.

Case Study

Waste Paper Creates Aisle-Space and Fire Violation

A follow-up to the Kaltech incident, led to an inspection in the Greenwich Village Section of Manhattan, New York City, of a printing company that occupied the first two floors of an eight-story building; the upper six floors were offices and residences. The inspection of the print shop revealed no serious violations, that is, until the inspector reached the hazardous waste storage area. The hazardous waste containers were stored on the loading dock, but were not accessible or even visible for inspection. All five of the 55-gallon drums of flammable wastes were totally blocked in by several tons of waste paper stored for recycling.

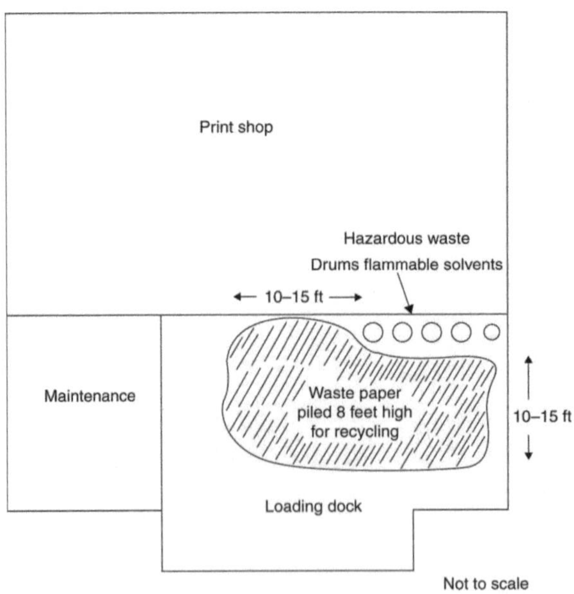

FIGURE 3.6 Print shop with aisle-space violations. (Drawing by author.)

Many aisle space violations under §265.173 are not considered serious violations, but this was no ordinary case. The seriousness of the violation stemmed from the facts that the building was for multiple use, the offices and residences were on floors above the potential fire hazard, and the flammable wastes were stored in drums surrounded by tons of loose papers. A fire in this area would have been catastrophic.

The owner was cited for the violation and was assessed a substantial penalty.

In this chapter, you learned more about U.S. hazardous waste management policy, studied the U.S. regulatory standards for categories of hazardous waste generators, hazardous waste transporters and treatment, storage and disposal facilities. You also read case studies on sham recycling of waste tires, storage of incompatible materials, the dangers of mixing incompatible wastes, universal waste confusion, a dry cleaner in New York City (local regulations supersede federal rules), bulging drums in storage, and waste paper creating aisle-space violations.

In the next chapter, you will learn about processes to treat hazardous wastes, including physical-chemical processes, biological processes, and thermal processes; land disposal of hazardous wastes, including deep-well injection, surface impoundments, waste piles; and land treatment facilities; also, hazardous waste storage facilities such as salt bed formations, underground mines or caves, concrete bunkers or vaults; and you will learn about hazardous waste landfills. You will also read a case study on issues with a RCRA-exempted steel drum cooperage (recyler/rebuilder).

EXERCISES

1. What are the main components of the federal hazardous waste management hierarchy?
   a. How does EPA define Pollution Prevention?
   b. How does EPA define Source Reduction?
   c. In your own words, define recycling
2. Under what circumstances may a nonregulated handler of hazardous waste generate hazardous waste and still be nonregulated?
3. The USEPA made the latest major changes to the Solid Waste Rules on what date? (answer September 25, 2015)
4. When do these changes take effect in EPA authorized states? (Answer On the date when each authorized state promulgates the state rules and regulations.
5. What are the hazardous waste generation limits for a Very Small Quantity Generator (VSQG) of hazardous waste?
6. What are the hazardous waste generation limits for a Small Quantity Generator (SQG) of hazardous waste?
7. Compare the regulatory requirements for a VSQG and a SQG of hazardous waste.
8. What are the hazardous waste generation limits for a Large Quantity Generator (LQG) of hazardous waste?

9. Compare the regulatory requirements for a SQG and a LQG of hazardous waste.
10. Do universal wastes have to be counted as part of the facility's hazardous waste generation?
    a. Under what circumstances do universal wastes get counted as regular hazardous waste?

REFERENCES

[CSB 02] Kaltech industries building explosion, Chemical Safety Board Digest, New York, New York, April 25, 2002, online at *http://www.csb.gov/assets/document/Kaltech_Digest.pdf*, (accessed January 2011).

[EPA 06] EPA Region 5 Scrap Tire Cleanup Guidebook (January, 2006). Online at *http://www.epa.gov/reg5rcra/wptdiv/solidwaste/tires/guidance/*, (accessed January 2011).

[EPA 11d] EPA Region 2 RCRA Self-Assessment Tools, online at *http://www.epa.gov/region2/capp/cip/rcra.htm*, (accessed January 2011).

[EPA 11e] EPA Hazardous Waste Manifest Form, online at *http://www.epa.gov/osw/hazard/transportation/manifest/pdf/newform.pdf*, (accessed January 2011).

[PPA 90a] Section 6602 (b) of the Pollution Prevention Act of 1990, online at *http://www.epa.gov/p2/pubs/p2policy/definitions.htm*, (accessed January 2011).

[PPA 90b] Pollution Prevention Act of 1990, U.S. Code Title 42, the Public Health and Welfare Chapter 133 §13032, online at *http://www.epa.gov/p2/pubs/p2policy/act1990.htm*, (accessed May 2011).

CHAPTER 4

HAZARDOUS WASTE TREATMENT AND DISPOSAL

In This Chapter

- Processes to treat hazardous wastes
- Land disposal of hazardous wastes
- Hazardous waste storage facilities
- Hazardous waste landfills
- Case study

Hazardous waste regulations apply to the storage, transportation, and treatment of hazardous waste, but in general terms; these regulations do not apply when the waste is discharged to an air pollution control system or a wastewater treatment system. That does not mean the waste ceases to be hazardous when discharged to another system—it means the waste has crossed a "regulatory boundary." Hazardous waste discharges to air pollution control devices or systems are regulated under the Clean Air Act (Chapter 6, "Air Pollution Control") and hazardous wastewater discharges to wastewater pollution control devices or systems are regulated under the Clean Water Act (Chapter 7, "Wastewater Management").

Following are basic descriptions of some technologies used to treat and dispose of hazardous waste under RCRA.

4.1 HAZARDOUS WASTE TREATMENT PROCESSES

After all possible means of source reduction, pollution prevention, recycling, reclamation, and reuse, the USEPA hazardous waste management hierarchy then requires the treatment of *all remaining hazardous wastes* to attain the treatment standards specified in the federal Land Disposal Restrictions Program (LDR) before their land disposal. Descriptions of the basic forms for hazardous waste treatment are as follows:

Numerous portions of this chapter have been quoted directly with permission from Dr. S. Amal Raj, Introduction to Environmental Science and Technology. Laxmi Publications Pvt. Ltd. 2008. 79-84.

Physical-Chemical Processes

The most common physical-chemical processes used to treat hazardous wastes are:

- Air stripping
- Ion exchange
- Adsorption
- Neutralization
- Precipitation
- Coagulation and flocculation
- Oxidation and reduction [Raj 08].

Air Stripping

Air stripping is a technology normally used to clean up contaminated hazardous waste sites. It is discussed in detail in Chapter 5, "Hazardous Waste Site Cleanup Technologies;" briefly, however, air stripping is the forcing of air through contaminated ground water or surface water to remove volatile chemicals.

Ion Exchange

Ion exchange is a chemical treatment process used to remove dissolved ionic species from contaminated aqueous streams. Ion exchange processes can achieve treatment of both anionic and cationic contaminants.

Soluble hazardous constituents that are amenable to treatment by ion exchange include arsenic, barium, cadmium, chromium, cyanide, lead, mercury, and silver. Ion exchange processes are often cost prohibitive for treatment of highly concentrated waste streams. Ion exchange is typically used as a polishing step after chemical precipitation. Stringent discharge limits cannot be met using ion exchange [Raj 08].

Adsorption

Many hazardous wastes contain organics which are refractory and are difficult to remove by conventional biological treatment processes. These materials can frequently be removed by adsorption on an active-solid surface. The most widely used adsorbent in environmental applications is carbon that has been processed to significantly increase internal surface area (activated carbon). Activated carbon is available in both granular and powdered form. Granular activated carbon is widely used for removal of a wide range of toxic organic compounds and heavy metals from groundwater and industrial waste streams. Activated carbon is often used as a polishing step in hazardous waste treatment, because if the waste stream is highly concentrated with contaminants that adsorb to the carbon, the activated carbon medium fills up quickly and the contaminants "break through" without adsorbing. Certain low molecular weight compounds like benzene (formula) and carbon tetrachloride (formula) are not amenable to carbon adsorption because they "break through" the carbon medium very quickly [Raj 08].

Neutralization

Acid-base reactions are among the most common chemical processes used in wastewater treatment. Neutralization of a waste involves addition of a chemical substance to change the pH to a more neutral level (6 – 8). The neutralization reactions are often exothermic and require systems similar to those pictured in Figure 4.1 to avoid excessively high temperatures, which could produce unsafe operating conditions and damage the process equipment [Raj 08].

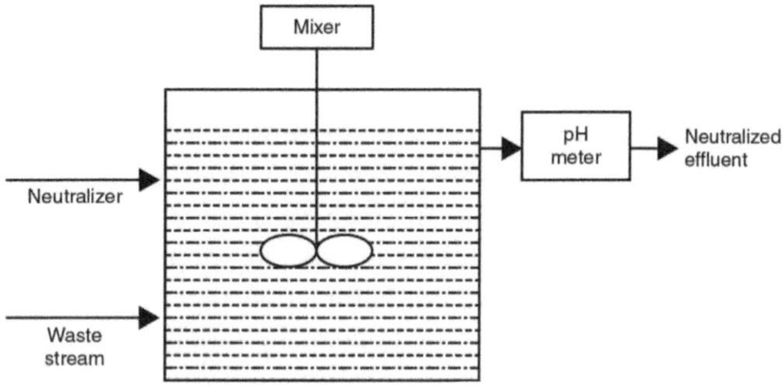

FIGURE 4.1 Neutralization process. (Permission granted courtesy of Laxmi Publications Pvt. Ltd.)

Precipitation

The undesirable heavy metals present in liquid waste streams can be removed by chemical precipitation. Metals are precipitated at varying pH levels, depending on the metal ion, resulting in the formulation of an insoluble salt. Hence, neutralization of an acidic waste stream can cause precipitation of heavy metals and allow them to be removed as a sludge residue by sedimentation, followed by filtration. The hydroxide of heavy metals is usually insoluble, so lime or caustic soda is commonly used to precipitate them [Raj 08].

Coagulation and Flocculation

Coagulation is the destabilization of particles in a liquid so that they can agglomerate to produce larger particles. Common coagulants that can be added are iron or aluminum salts that react with alkalinity to form metal hydroxides.

Flocculation is the coagulation or agglomeration of coagulated particles (normally follows coagulation) and causes coagulated particles to form larger flocs which can then settle (or float) faster.

Oxidation and Reduction

The chemical processes of oxidation and reduction can be used to convert toxic pollutants to harmless or less toxic substances. Oxidation is a chemical reaction in which the valence of a compound increases from the loss of electrons. Reduction is the opposite reaction, where valence of a compound decreases from gaining electrons. Chemical reactions involving both oxidation and reduction are called redox reactions.

An example of a reduction reaction is as follows:

Hexavalent chrome is highly toxic and its presence in a waste requires careful management to avoid harm to human health and the environment. When hexavalent chrome is reduced to trivalent chrome, it can be precipitated as chromic hydroxide, as shown in the following reaction that uses sulfur dioxide and lime:

$$SO_2 + H_2O \rightarrow H_2SO_4$$
$$2CrO_3 + 3H_2SO_3 \rightarrow Cr_2(SO_3) + 3H_2O$$
$$Cr_2(SO_4)_3 + 3Ca(OH)_2 \rightarrow 2Cr(OH)_3 + 3CaSO_4$$

The above reduction of hexavalent chromium to the trivalent state through techniques similar to the above produces a chromium-containing compound that is less toxic and more acceptable for subsequent recovery or final disposal [Raj 08].

Oxidation and Reduction reactions can also be used to change halogenated compounds to non-halogenated compounds.

Biological Processes

Biological degradation of hazardous organic substances is a viable approach to waste management. Common processes are those originally utilized in treating municipal wastewater, based on aerobic or anaerobic bacteria. In-situ treatment of contaminated soils can be performed biologically, provided the conditions in the soil are amenable to the biota needed for the treatment. Cultures used in biological degradation processes can be native (indigenous) microbes, selectively adapted microbes, or genetically modified microorganisms. Cultures of bacteria are often borrowed from wastewater treatment systems treating similar contaminants and are introduced as a "seed" or "starter solution" so the bacteria can adapt and grow to treat the targeted contaminants.

Biological treatment processes are applied to gaseous, aqueous, and solid phases containing biodegradable organic compounds and inorganic ions such as nitrate, ammonia, sulfate, and phosphate. Microorganisms involved in biological hazardous waste treatment processes include:

- Bacteria
- Fungi
- Protozoa
- Algae

The most active and diverse group are the bacteria. For microorganisms to degrade hazardous contaminants, conditions that promote their growth and reproduction must be maintained. Required for this are sources of energy, carbon, nitrogen, phosphorous, and trace micronutrients, along with proper environmental conditions such as temperature, pH, and moisture. In addition, the success of biological treatment relies on the biodegradability of the contaminants of interest. Major factors that affect the biodegradability of a specific contaminant are:

- Presence of an appropriate microbial culture
- Chemical structure of the contaminant
- Physical characteristics of the contaminant

Each of these factors must be considered in the choice of the biological reactor configuration to promote the desired degradation reactions [Raj 08].

Thermal Treatment

Thermal treatment applies high temperature to convert hazardous waste to forms that are significantly less toxic, have lower volume, and are more easily disposed of. In general, two types of thermal technologies are used. They are:

- Incineration
- Pyrolysis

Incineration

Incineration involves the combustion of waste in the presence of oxygen.

- Incinerators—used primarily for waste destruction
- Boilers and Industrial Furnaces (BIFs)—used primarily for energy and material recovery

Although incinerators are used to burn hazardous waste primarily for waste destruction/treatment purposes, the additional benefits of energy or material recovery can occur in certain circumstances.

When performed properly, incineration destroys the toxic organic constituents in hazardous waste and reduces the volume of the waste. There are many types of hazardous waste incinerators including:

- Rotary kilns
- Fluidized bed units
- Liquid injection units
- Fixed hearth units

Rotary Kilns

The rotary-kiln incinerator is used by municipalities for municipal solid waste incineration and by large industrial plants for solid waste or hazardous waste incineration. Rotary kilns are also used to produce cement and lightweight aggregate. These types of incinerators are used for hazardous waste destruction because the wastes are exposed to high temperatures for relatively long residence times.

The rotary kiln has two chambers: a primary chamber and secondary chamber.

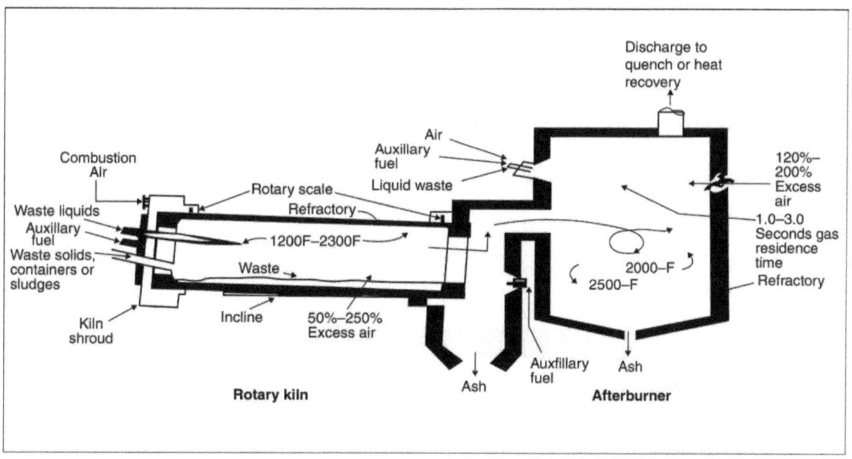

FIGURE 4.2 Typical rotary kiln incinerator. (Adapted from EPA-530-R-94-014.)

The primary chamber is a long inclined cylinder, lined with a refractory material, such as brick. The cylinder rotates while the material to be incinerated tumbles down the inclined tube. During this tumbling/heating process, solids and liquids are converted to gases through various reactions, including destructive partial combustion, distillation, and volatilization.

In the secondary chamber, the combustion of the gaseous phase is completed, with the assistance of additional fuel, as necessary. The combustion process is also aided by additional combustion air (draft), by using a fan, flue gas stack, steam jet, or a combination of these. Any incombustible solids (clinkers) spill out of the lower end of the cylinder, and any remaining gases and particles are often burned in an afterburner. Air pollution devices can be used to treat any of these remaining gases. The particulate materials from hazardous waste incinerators are usually cooled down and filtered through a "baghouse," where the particulates are trapped in filter bags. The materials recovered in the baghouses are often recycled through the kilns to be incorporated into the product for beneficial reuse [WIKI 11c].

Fluidized Bed Incinerators

In a fluidized bed incinerator, the bed is actually a mixed and churning bed of sand at the bottom of the burner, created by pumping large volumes of air through the sand until the sand particles physically separate. This creates a fluid-like environment, in which fuel and waste can be introduced. These fluidized beds promote optimum contact between the solids, liquids and gases, and have the advantages of:

- Very high mixing of particles of waste and fuel during combustion phase
- High frequency of particle collision
- Very high surface area contact between solids and fluids [WIKI 11c].

The fluidized bed incinerator is very effective for destruction of liquid wastes.

Liquid Injection Incinerators

In a liquid injection incinerator, liquid wastes are atomized into a combustion chamber lined with a refractory material, where they are burned. If the waste has an adequate heating value, it can be combusted alone, but if the heating value is too low, additional fuel must be added. The temperature achieved in combustion ranges from 1,200 to 3,000 degrees Fahrenheit, and the residence time of the waste in the incinerator is very rapid, compared with most other types of incineration, ranging from 0.5 to 2 seconds.

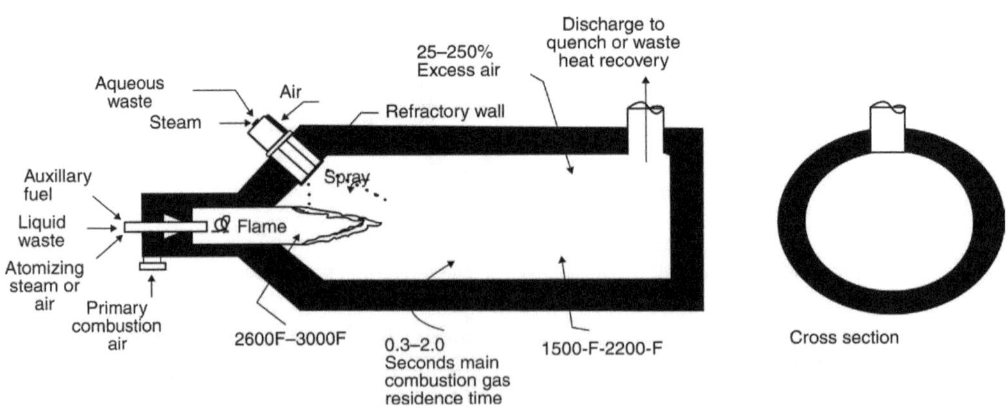

FIGURE 4.3 Typical liquid injection incinerator. (Adapted from EPA-530-R-94-014.)

If the energy content of the waste is not high enough to maintain adequate ignition and incineration temperatures, a supplemental fuel, such as natural gas or fuel oil, may be pumped into the combustion chamber through a separate nozzle or mixed in the same fuel feed to augment the ignition potential of the waste mix. Air necessary for the combustion is provided by a fan [EPA 98].

Fixed Hearth Incinerators

The fixed hearth incinerator uses very simple and basic principles of incineration technology and is used primarily for municipal solid waste. The waste is placed on a fixed grate in a chamber lined with refractory material and is introduced to a flame, which creates three separate waste streams. Combustion gases are released through a stack through pollution control equipment; any incombustible solids (clinkers) are removed through another opening, and the ashes fall through the fixed grate for disposal [WIKI 11e]. This type of incinerator has the advantage of being able to handle solid phase wastes that cannot be handled in liquid incinerators.

NOTE *Because metals will not combust, incineration is not an effective method for treating metal-bearing hazardous wastes.*

Boilers and Industrial Furnaces (BIFs)

BIFs are categorized separately by the EPA because they typically treat wastes as a secondary benefit. EPA gives them this special designation because the primary benefit of BIFs is either significant energy recovery or material recovery.

Boilers use flame combustion in a controlled manner to recover energy or materials from the waste.

Industrial furnaces use thermal treatment or incineration to recover energy or materials as an integral part of a manufacturing process. Two good examples of industrial furnaces are cement kilns and aggregate kilns, where the energy generated from burning waste is used to heat the cement or aggregate in the process, reducing the need for use of virgin fuels in those processes. Other examples of industrial furnaces are: smelters, refining furnaces, blast furnaces, halogen acid furnaces, and methane reforming furnaces [EPA 11f].

NOTE *The BIF terminology can be confusing in that cement kilns, lime kilns, aggregate kilns, and phosphate kilns are usually rotary kiln incinerators but are defined by EPA as industrial furnaces. EPA gave them this regulatory distinction in order to allow energy and material recovery potential, with waste treatment being a secondary benefit.*

Hazardous waste streams containing significant quantities of organics and minimal amounts of metals and solids are suitable for incineration. Wastes containing high concentrations of halogenated compounds or volatile metals are unsuitable for incineration. Incineration is an oxidation process in which wastes are heated to a high temperature for a minimum prescribed "residence" time in the presence of oxygen. Organics within the waste are converted to CO_2 and water, nitrogen and sulfur are oxidized to inorganic gases, halogens are converted to acidic gases and salts, and metals are oxidized and precipitated into ash or volatilized. Ash is an inherent byproduct of incineration that must be collected and either disposed in a landfill or beneficially reused in concrete production [Raj 08].

NOTE *The air emissions from all thermal destruction processes used to treat hazardous wastes are subject to the Clean Air Act, which is discussed in Chapter 6.*

Several different incinerator configurations used for the treatment of hazardous waste and the operating temperatures are given in Table 4.1.

TABLE 4.1 Typical incinerator operating conditions

Incinerator Type	Temperature Range (°C)
Liquid Injection	1000 - 1700
Rotary Kiln	820 - 1600
Fluidized Bed	760 - 980
Multiple Hearth	720 - 980

Source: Permission granted courtesy of Laxmi Publications Pvt. Ltd.

The choice of an appropriate incinerator configuration depends on the characteristics and volume of the waste, incinerator availability, and cost, the details of which are beyond the scope of this text.

! TIP *Resource Conservation and Recovery Act (RCRA) regulations governing incinerators can be found at 40 CFR Part 264/265, Subpart O-Incinerators. Also 40 CFR 266 govern Boilers and Industrial Furnaces (BIFs)*

Pyrolysis

Pyrolysis is the thermal decomposition of molecules in the absence of oxygen.

"Pyrolysis is applicable for the treatment of wastes that are not amenable to conventional incineration. The major advantage associated with pyrolytic processes for hazardous waste treatment is more efficient energy recovery compared with incineration. Major limitations of this process include the requirement of auxiliary heating during the endothermic stage, long residence times compared to incineration units, potentially hazardous products in the gaseous emissions, and potentially hazardous leachate residues resulting from treatment of metals and salt-bearing wastes. Consequently, pyrolysis is currently not applied for the treatment of hazardous waste nearly as often as incineration" [Raj 08].

Precious Metals Recovery

Precious metals reclamation is the recycling and recovery of precious metals (i.e., gold, silver, platinum, palladium, iridium, osmium, rhodium, and ruthenium) from hazardous waste. Because these materials will be handled protectively as valuable commodities with significant economic value, generators, transporters, and facilities that store such recyclable materials are subject to reduced requirements.

Although this segment could be discussed in the recycling section of Chapter 8, "Solid Waste Management," it is appropriate to discuss the recovery of precious metals in this chapter, because many of the precious metals are recovered from hazardous wastes; for example, photographic processes (silver), and used electronics (gold, silver, platinum, palladium, ruthenium and iridium).

Silver can be recovered using an electrolytic process in which crude silver is dissolved at the anode and refined silver or copper is deposited at the cathode. This is a common method in recovery of silver from fixer/developer solutions from photographic and x-ray development.

Although copper (not a precious metal) can be refined using a melting process, the purest copper is obtained by an electrolytic process, using a piece of less-than-pure copper as the anode and a thin sheet of pure copper as the cathode. The electrolyte is an acidic solution of copper sulfate. Electricity is passed through the cell, and copper is dissolved from the anode and deposited on the cathode.

Unlike the base metals, the remainder of the precious metals does not oxidize or react chemically, so when they are heated at high temperatures, the precious metals remain apart and the other metals react, forming slags or other compounds [WIKI 11e].

The relaxed regulatory requirements for precious metals recovery make this effort attractive to many companies, in that if they meet certain requirements, they need only to obtain an EPA Identification Number, meet certain record-keeping requirements, and comply with the Land Disposal Restriction notification requirements.

4.2 LAND DISPOSAL

The term land disposal simply means depositing wastes in or on the land. The following technologies are employed as land disposal methods:

- Deep-well injection
- Surface impoundments
- Waste piles
- Land treatment units
- Landfill (See Section 4.4 below)

Deep-Well Injection

Underground (deep-well) injection involves using specially designed wells to inject liquid hazardous waste into deep earth strata containing non-potable water. In this method, a wide variety of hazardous waste liquids are pumped underground into deep permeable rock formations that are separated from freshwater aquifers by impermeable layers of rock above, below, and lateral to the waste layer. The depth of the injection ranges from about 300 to 2500 meters (900 - 7,500 feet), and varies according to the geographical factors of the area [Raj 08].

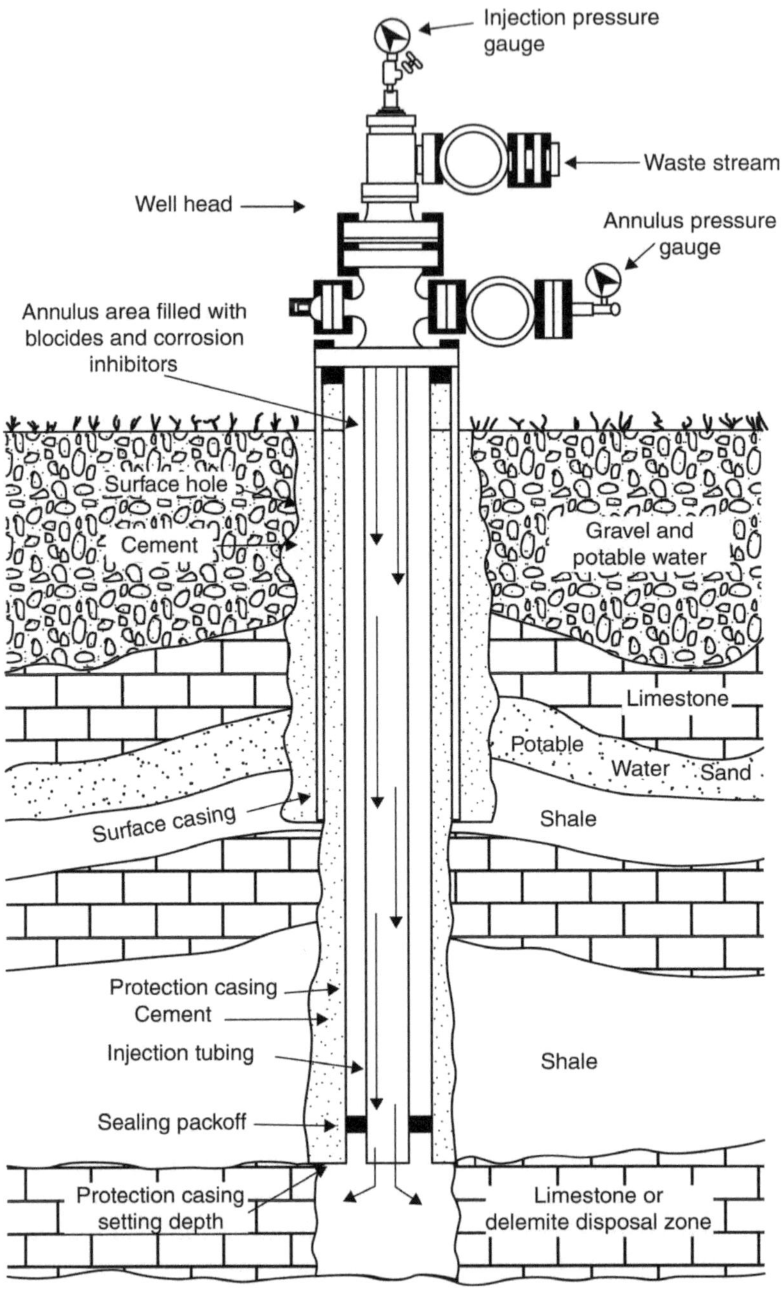

FIGURE 4.4 Hazardous-waste-disposal well. (Permission granted courtesy of Laxmi Publications Pvt. Ltd.)

The cross-section of a typical injection well is shown in Figure 4.4. The particulate matter present in the liquid should be removed to prevent plugging of the injection equipment. The deep well must be constructed such that potable water zones are isolated and protected. For hazardous waste liquids to be deep-well injected, the following guidelines must apply:

- Must be compatible with all other waste previously injected
- Must be biologically inactive
- Must be noncorrosive
- Must be difficult to be treated by other methods.

Thus, the method should be used only for those liquid wastes with no other feasible management options [Raj 08].

Surface Impoundments

"Surface impoundments are natural topographic depressions, man-made excavations, or diked areas formed primarily of earthen materials used for temporary storage or treatment of liquid hazardous waste. Examples include holding, storage, settling, aeration pits, ponds, and lagoons. Hazardous waste surface impoundments are required to be constructed with a double liner system, a leachate collection and removal system (LCRS), and a leak detection system. To ensure proper installation and construction, regulations require the unit to have and follow a construction quality assurance (CQA) program. The regulations also outline monitoring, inspection, response action, and closure requirements (see 40 CFR Parts 264/265, Subpart K)" [EPA 11g].

Waste Piles

"Waste Piles are non-containerized piles of solid, non-liquid hazardous waste that are used for temporary storage or treatment. In addition to the standard double liner and leachate collection and removal systems (LCRS), waste piles are required to have a second LCRS above the top liner. Waste piles must also have run on and runoff controls, be managed to prevent wind dispersal of waste, and are subject to inspection, monitoring, and release response requirements. When closing a waste pile, all waste residue and contaminated soils and equipment must be removed or decontaminated (see 40 CFR Parts 264/265, Subpart L)" [EPA 11g].

Land Treatment Units

"Land Treatment Units use naturally occurring soil microbes and sunlight to treat hazardous waste. This is accomplished by applying the hazardous waste directly on the soil surface or incorporating it into the upper layers of the soil in order to degrade, transform, or immobilize the hazardous constituents. Land treatment units rely upon the physical, chemical, and biological processes occurring in the topsoil layers to contain the waste. Because of this, the units are not required to have liner systems or leachate collection and removal systems (LCRS). Before hazardous waste can be placed in a land treatment unit, operators must complete a treatment demonstration to demonstrate the unit's effectiveness and ability to treat the hazardous waste. Once operational, operators must monitor the unit (unsaturated zone monitoring) to ensure that all hazardous constituents are being treated adequately. Unit closure consists primarily of placing a vegetative cover over the unit and certifying that hazardous constituent levels in the treatment zone do not exceed background levels (see 40 CFR Parts 264/265, Subpart M)" [EPA 11g].

4.3 HAZARDOUS WASTE STORAGE UNITS

For wastes that have no known treatment technology, such as low level radioactive wastes mixed with hazardous wastes, the USEPA recognizes the storage of these wastes as an acceptable alternative until the half-lives of these wastes have expired, or a treatment alternative is developed.

Examples of theses storage facilities are:

Salt Dome Formations, Salt Bed Formations, Underground Mines, and Underground Caves

"Salt Dome Formations, Salt Bed Formations, Underground Mines, and Underground Caves are geologic repositories. Because these units vary greatly, they are subject to environmental performance standards, not prescribed technology-based standards (e.g., liners, leachate collection systems, leak detection systems). The standards require that these miscellaneous units must be located, designed, constructed, operated, maintained, and closed in a manner that ensures the protection of human health and the environment (See 40 CFR Part 264 Subpart X—Miscellaneous Units)" [EPA 11g].

Concrete Bunker or Vault

Concrete bunkers and vaults are repositories made of concrete used to store hazardous wastes over long periods of time. These facilities are most commonly used for radioactive wastes, or mixed wastes (hazardous wastes mixed with low-level radioactive wastes) for which there is no technology yet available for treatment.

4.4 HAZARDOUS WASTE LANDFILLS

Sanitary landfills were developed for municipal solid waste disposal to replace existing open dumps. New secure hazardous waste landfills are used to bury only treated and stabilized residuals from hazardous waste treatment in artificially lined depressions. Secure landfills for hazardous waste residuals are now equipped with double liners, leak detection, leachate collection, monitoring, and leak detection systems. The schematic representation of a secure double lined landfill system is shown in Figure 4.5 [EPA 11g].

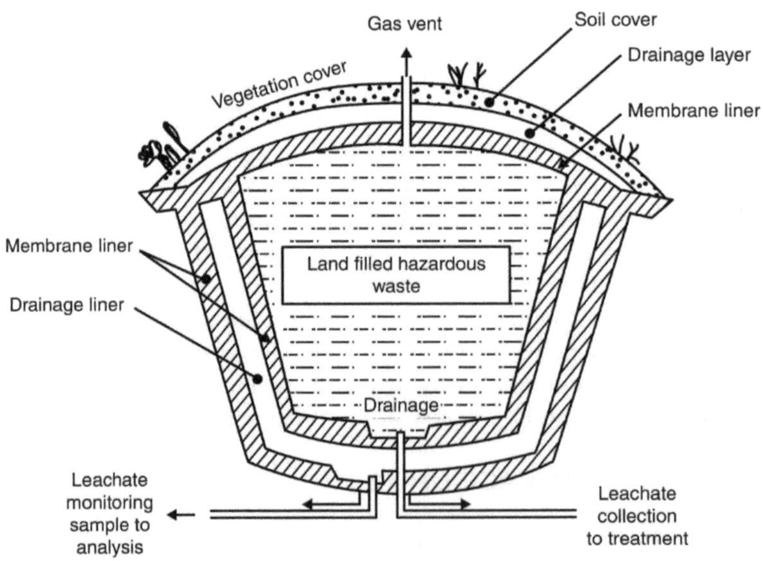

FIGURE 4.5 Double-lined landfill after closure. (Permission granted courtesy of Laxmi Publications Pvt. Ltd.)

The following factors are to be considered while selecting the site for any landfill:

- All applicable siting criteria for the host state and communities must be followed
- Surface runoff should be intercepted and diverted
- The integrity of top soil covers should be maintained
- Surface erosion should be prevented
- Artesian pressure should be released
- Ground water should not move laterally
- Liners should be properly chosen and installed
- Leachate collection and removal systems are provided
- Perpetual post-closure maintenance is provided
- Environmental monitoring around the site is conducted
- Security around the site is maintained in perpetuity [Raj 08]

Case Study

Issues with Exempted Steel Drum Cooperage

An interesting feature of the hazardous waste regulations is that the residues from empty containers that held hazardous waste are exempt from regulation, regardless of the type or concentration of the waste. This makes it very interesting to conduct a hazardous waste inspection at a steel drum cooperage that reclaims or refurbishes "empty" steel drums.

The word cooperage originally described a facility that built casks or barrels. The term was later used to describe facilities that cleaned and refurbished steel drums, regardless of whether they had been used for food products, hazardous materials, or even hazardous wastes.

In the hazardous waste regulations, "RCRA empty" means a container is only RCRA empty, per 261.7, if after pouring, pumping, and aspirating, it then holds less than one inch or a certain percentage of the container contents of residue. A facility cleaning thousands of "RCRA empty" steel drums for reclaiming or refurbishment can potentially end up with hundreds or even thousands of gallons of "exempt" hazardous wastes. This loophole in the regulations caused an environmental disaster in upstate New York, where the owner/operator of a cooperage discharged all of the materials taken from all of his steel drums into a pit, without treating the wastes at all.

This facility was caught dumping these wastes in the early 1980s, and the owner was required to clean up the pollution caused by his actions. Substantial penalties were levied, and the owner passed away before the cleanup was completed by the USEPA.

In this chapter, you learned about processes to treat hazardous wastes, land disposal of hazardous wastes, hazardous waste storage facilities, and hazardous waste landfills. You also read a case study on issues with a RCRA exempted steel drum cooperage (recycler / rebuilder).

In the next chapter, you will learn about treatment technologies for contaminated hazardous waste sites, containment technologies for contaminated sites, and a few innovative technologies for contaminated sites. You will also read case studies involving a compressed gas facility going out of business, a poorly operated mercury recovery facility creating a cleanup site, picric acid in an office building, the dangers of methamphetamine labs, and an abandoned fireworks factory.

EXERCISES

1. What is the regulatory status of hazardous wastes when they are discharged to wastewater systems or air pollution systems?
2. Name the three main categories of technologies for hazardous waste management.
3. At what point in a treatment system is ion exchange normally employed?
4. What treatment is often used for hard-to-remove refractory organic compounds?
5. What is an effective way to remove heavy metals from a liquid waste stream?
6. How do operators of wastewater treatment systems get their systems started for particular waste streams?
7. What are the two types of thermal treatment?
8. What is the key difference between the two?
9. Explain the purpose of the two chambers in a rotary kiln incinerator.
10. What types of wastes are most amenable to fluidized bed incinerators?
11. What types of wastes are not treated effectively by incineration?
12. Name five types of industrial furnaces.
13. Why is mercury easier to recover than many other metals?
14. Aside from its recovery value, why is precious metals recovery from hazardous waste attractive to many companies?
15. Name three forms of hazardous waste land disposal.
16. Name three types of hazardous waste long-term storage facilities.

REFERENCES

[EPA 11f] USEPA Hazardous Waste—Treatment & Disposal, online at *http://www.epa.gov/osw/hazard/tsd/td/combustion.htm*, (accessed May 2011).

[EPA 11g] USEPA - Hazardous Waste Land Disposal Units (LDUs), online at *http://www.epa.gov/osw/hazard/tsd/td/ldu/*, (accessed May 2011).

[EPA 98] USEPA Solid Waste and Emergency Response EPA-542-R-97-012 March 1998, online at *http://www.epa.gov/tio/download/remed/incpdf/incin.pdf*, (accessed May 2011).

[Raj 08] Raj, S. Amal. 2008. Introduction to environmental science and technology. 79-84. Laxmi Publications Pvt, Ltd.

[WIKI 11c] Wikipedia.com, online at *http://en.wikipedia.org/wiki/Incineration#Rotary-kiln*, (accessed May 2011).

[WIKI 11d] Wikipedia.com, online at *http://en.wikipedia.org/wiki/Fluidized_bed*, (accessed May 2011). [WIKI 11e] Wikipedia.com, online at *http://en.wikipedia.org/wiki/Synthesis_of_precious_metals*, (accessed March 2011).

CHAPTER 5

Hazardous Waste Site Clean-up Technologies

In This Chapter

- Treatment technologies for contaminated hazardous waste sites
- Containment technologies for contaminated sites
- Innovative technologies for contaminated sites
- Case studies

As discussed in Chapter 1, "Introduction to Hazardous Waste," the Toxic Waste Act—also called Superfund—is a federal law of the United States that was created to enforce the investigation and cleanup of deserted hazardous waste sites. The contamination from these deserted sites is often below ground, and the hazardous wastes and other pollutants are not as easily accessible as wastes generated at typical industries. Therefore, the technologies used to treat or otherwise manage these wastes are different from conventional technologies. Some technologies have been developed to treat waste where it is found; other technologies have been developed to treat contaminated media once it is brought to the surface. Additionally, for the sites where neither type of these technologies is available, there are management methods to keep the waste from migrating off site. Several examples of each technology are discussed in this chapter.

A number of the descriptions of the technologies discussed in this chapter were adapted from a USEPA Superfund Web site, USEPA Superfund Remediation Technologies, at

http://www.epa.gov/superfund/remedytech/remed.htm.

This Web page provides individual links to further information about each of the technologies listed below.

5.1 TREATMENT TECHNOLOGIES FOR CONTAMINATED HAZARDOUS WASTE SITES

Although all of the technologies discussed in Chapter 4, "Hazardous Waste Treatment and Disposal," are sometimes used for cleanup or "remediation" of inactive hazardous waste sites, there are several additional, specialized technologies that have been employed or are currently in use almost exclusively at cleanup sites. These specialized technologies are necessary when the contamination is trapped underground in groundwater or rock and soil matrices and cannot be reached for conventional treatment.

This chapter is devoted to describing a number of cleanup technologies that are available and are used in the United States today. This is not necessarily a complete list, but most of the popular technologies are represented.

The technologies discussed in this chapter are grouped into two general classifications of treatment for hazardous waste cleanup sites: *in-situ* treatment may be defined as the treatment of contaminants prior to their removal from the soil and groundwater; *ex-situ* treatment involves removing the contaminant from its existing location and treating it above the surface.

In-Situ (On-Site) Treatment of Contaminated Hazardous Waste Sites

In-Situ Air Entrainment

One method used to treat hazardous contaminants in the soil or groundwater is to pump air into the contaminated zones. The two methods, air stripping and air sparging, are listed below.

- ***Air stripping*** is the forcing of air through contaminated groundwater or surface water to remove harmful chemicals. This technology is most effective when the groundwater is easily accessible and the contaminants are volatile and/or semi-volatile and easily susceptible to treatment by exposure to air molecules. Air stripping is ineffective if the contaminants are bound to the particles of rock and soil.

A schematic of air stripping of media in a vessel is depicted in Figure 5.1.

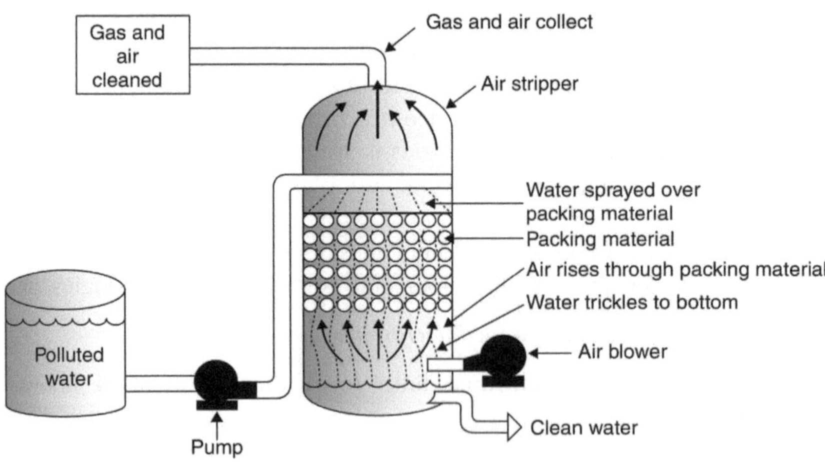

FIGURE 5.1 Air stripper. (Adapted from EPA 542-F-01-016.)

- ***Air sparging*** involves the injection of air or oxygen through a contaminated soil zone. Injected air traverses in channels through the soil column both horizontally and vertically. This creates an underground stripper that removes volatile and semi-volatile organic compounds by volatilizing these contaminants. Air is injected to help flush the contaminants into the *vadose* (unsaturated) zone. Sometimes oxygen is added to the contaminated soils to optimize biodegradation both above and below the water table. Air sparging in combination with soil vapor extraction is illustrated in Figure 5.2.

Soil Vapor Extraction

NOTE

Soil Vapor Extraction (SVE) can be used as an in-situ or ex-situ treatment method and can be used in combination with many other types of treatments. SVE is used to remove contaminants from unsaturated (vadose) zone soils. A vacuum is applied to the soil to facilitate the controlled flow of air, removing volatile and some semi volatile organic contaminants from the soil. SVE usually is performed in-situ; however, in some cases, it can be used as an ex-situ technology. (Any vapor treated after removal is treated in an ex-situ process.)

Figure 5.2 is an illustration of SVE combined with air sparging.

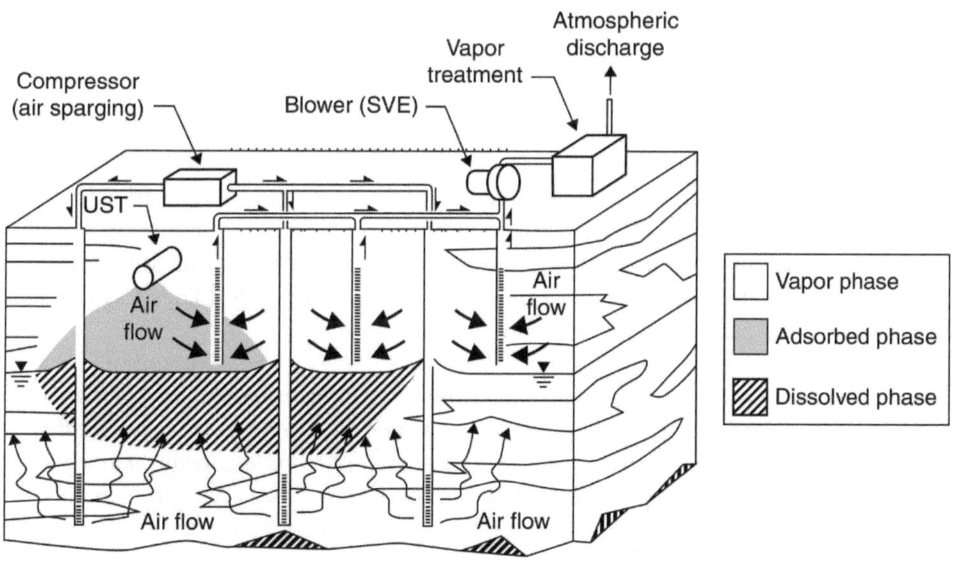

FIGURE 5.2 Air sparging using SVE. (Adapted from EPA 510-B-95-007.)

In-Situ Bioremediation

Bioremediation is a broad term that connotes using microbes to clean up harmful chemicals in contaminated media. This technology is possible only if the proper amounts of oxygen and nutrients are available at appropriate temperatures. Bioremediation is most effective when the contaminants are consistent, and the temperature and other environmental factors are steady-state. In-situ bioremediation techniques create and stimulate a favorable environment for microorganisms to grow and use various contaminants as a food and/or energy source. The treatment is effective only when the proper combination of oxygen, nutrients, and moisture are provided, attendant with temperature and pH control. Sometimes, microorganisms that have been adapted for degradation of specific contaminants are applied to enhance the process. Two examples of bioremediation are given below:

- ***Bioremediation of chlorinated solvents*** "uses microorganisms to break down contaminants by using them as a food source or by cometabolizing them with a food source. Aerobic processes require an oxygen source, and the end products typically are carbon dioxide and water. Anaerobic processes are conducted in the absence of oxygen, and the end products can include methane, hydrogen gas, sulfide, elemental sulfur, and dinitrogen (N_2) gas" [EPA 11h].
- ***Chemical oxidation*** involves pumping chemicals called oxidants down into wells to destroy pollution in soil and groundwater. Oxidants react with harmful chemicals to reduce them to harmless ones, like water and carbon dioxide. Chemical oxidation is used to treat many chemicals such as fuels, solvents and chemicals [EPA 11h].

In-Situ Physical Treatment

There are a number of physical treatment technologies used for cleanup:

- ***Soil excavation*** is arguably not a treatment; however, digging up polluted soils in order for them to be placed in a permitted landfill is necessary when no other effective or practical form of treatment is available.
- ***Pump and treat*** is a method that uses pumps to bring contaminated liquids up to the surface, where the liquids are either treated and discharged or taken off site for treatment.
- ***Fracturing*** uses a liquid or air to fracture rocks or dense soils in order to enable other types of treatment. As with soil excavation, fracturing is not a treatment in and of itself.
- ***In-situ flushing*** is the process of pumping uncontaminated, harmless liquids into wells to push contaminated liquids underground toward recovery wells. (These liquids can then be pumped to the surface for treatment in an ex-situ process.)

NOTE

Permitting is not usually an issue for hazardous waste cleanups because they are not ordinarily required to obtain a hazardous waste permit. Facilities set up to treat wastes generated at hazardous waste clean-up sites are exempt from hazardous waste permitting provided they meet the Applicable or Relevant and Appropriate Requirements (ARARs). These requirements are spelled out in the Superfund Law (CERCLA), which was discussed in Chapter 1,"A Brief History of Hazardous Waste".

In-Situ Thermal Treatment

This treatment method uses several means of thermal elevation of soils to destroy contaminated soils and liquids or to push them toward collection wells. Often, combinations of thermal treatment that include thermal conduction, electrical resistance, steam, hot air, hot water, or radio frequency, are used. Thermal treatment is very effective on non-aqueous phase liquids (NAPL), which do not migrate or dissolve easily in groundwater. (Materials removed by this process are then treated in an ex-situ process such as wastewater treatment.)

- ***Multiphase extraction*** uses a powerful vacuum system to remove subsurface contaminants such as groundwater, vapors, and petroleum product. This lowers the water table around the well, exposing more of the contamination in the formation. These contaminants in the exposed vadose zone are then more accessible to SVE. Once above ground, the extracted vapors or liquid-phase organics and groundwater are separated and treated ex-situ, usually by wastewater treatment.
- ***Phytoremediation*** uses plants and trees to treat contaminated soils. The roots of the plants take up the contaminated liquids and use them as nutrients. This process converts some of those chemicals into less hazardous ones. The roots of the plants also significantly reduce erosion caused by wind and precipitation. This type of treatment works only in climates appropriate for the plants and on chemicals that are not toxic to the plants.
- ***Monitored natural attenuation*** uses natural conditions and naturally occurring bacteria that are already present to clean up contaminants. Conditions are monitored to make sure they are amenable for the natural treatment to occur.
- ***Permeable reactive barriers*** are underground walls filled with permeable, reactive material, arranged in a long narrow trench. The reactive material treats the contaminated liquids as they migrate through the wall, sometimes pushed through the barrier by pumping nonhazardous liquids into wells that are located along one side of the wall. An illustration of a permeable reactive barrier is shown in Figure 5.3.

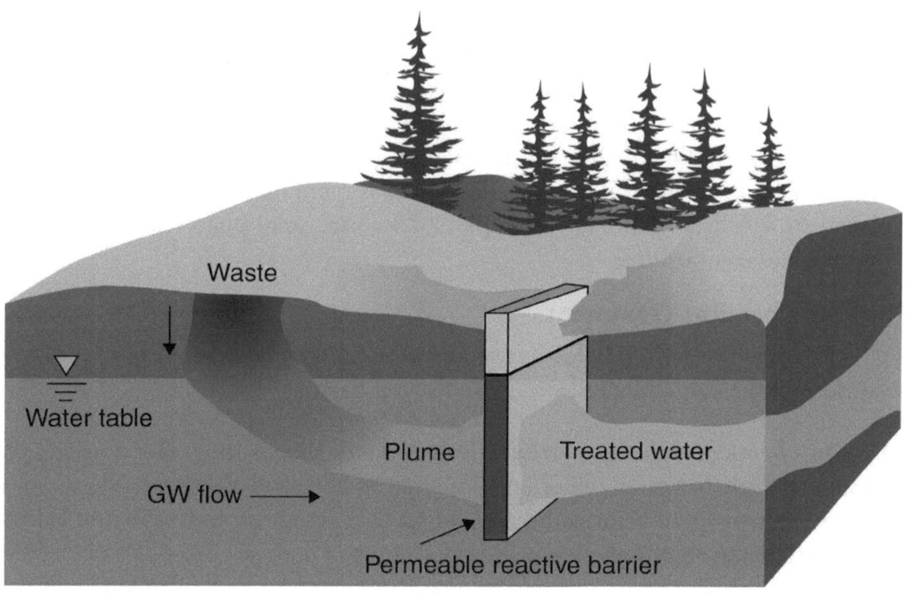

FIGURE 5.3 Permeable reactive barrier. (Adapted from EPA/600/R-98/125.)

- "***Circulating wells (CWs)*** provide a technique for subsurface remediation by creating a three-dimensional circulation pattern of the ground water. Groundwater is drawn into a well through one screened section and is pumped through the well to a second screened section where it is reintroduced to the aquifer. The flow direction through the well can be specified as either upward or downward to accommodate site-specific conditions. Because groundwater is not pumped above ground, pumping costs and permitting issues are reduced and eliminated, respectively. Also, the problems associated with storage and discharge are removed. In addition to groundwater treatment, CW systems can provide simultaneous vadose zone treatment in the form of bioventing or soil vapor extraction. CW systems can provide treatment inside the well, in the aquifer, or a combination of both. For effective in-well treatment, the contaminants must be adequately soluble and mobile so they can be transported by the circulating ground water. Because CW systems provide a wide range of treatment options, they provide some degree of flexibility to a remediation effort" [EPA 11h].

- "***Electrokinetics*** relies upon application of a low-intensity direct current through the soil between ceramic electrodes that are divided into a cathode array and an anode array. This mobilizes charged species, causing ions and water to move toward the electrodes. Metal ions, ammonium ions, and positively charged organic compounds move toward the cathode. Anions such as chloride, cyanide, fluoride, nitrate, and negatively charged organic compounds move toward the anode. Removal of contaminants at the electrode may be accomplished by several means, among which are: electroplating at the electrode; precipitation or co-precipitation at the electrode; pumping of water near the electrode; or complexing with ion exchange resins" [EPA 11h].

Ex-Situ (Off-Site) Remediation

Examples of off-site remediation (contaminants or contaminated media brought to the surface for treatment at the cleanup site or at a facility removed from the site) include:

Ex-Situ Bioremediation

- ***Slurry-phase bioremediation*** mixes excavated contaminated soils in water to form a slurry. This slurry keeps solids suspended and microorganisms in contact with the soil contaminants, thus facilitating treatment.
- ***Solid-phase bioremediation*** mixes added water and nutrients with soils, which are then placed in a cell or building. Examples of ex-situ, solid-phase bioremediation are land farming and composting.

Ex-Situ Physical/Chemical Treatment

Each of the methods listed below is used to treat contaminants after they have been brought to the surface of the contaminated site.

- ***Ex-situ chemical dehalogenation*** involves the digging, sifting, and crushing of soils contaminated with halogens (chlorine, bromine, iodine, and fluorine). These soils are then placed in an on-site reactor with heat and specific chemicals to reduce the halogen content or to change the chemicals to less hazardous ones. There are two types of chemical dehalogenation:
  - "*Glycolate dehalogenation* adds a combination of two chemicals known as alkaline polyethylene glycol (APEG) to soil in a reactor similar to the one described above. During mixing and heating, one chemical combines with the halogens to form a nontoxic salt. The other replaces the halogens to form other nontoxic chemicals. The heat in the reactor can cause some of the chemicals to evaporate. The gases are treated by air pollution control equipment at the site" [EPA 11h].

- *Base-catalyzed decomposition* is a process in which contaminated soil is excavated and "sodium bicarbonate is added to the soil in the reactor. The sodium bicarbonate allows the harmful chemicals in the soil to evaporate at a low temperature. Once the chemicals evaporate, the cleaned soil can be returned to the site" [EPA 11h].

- "***Solvent extraction*** uses an organic solvent to separate organic and metal contaminants from soil. First, the organic solvent is mixed with contaminated soil in an extraction unit. The extracted solution then is passed through a separator, where the contaminants and extractant are separated from the soil. Organically bound metals may be extracted along with the target organic contaminants" [EPA 11h]. (Although the extraction is sometimes an in-situ process, any treatment of these extracted materials is classified as ex-situ.)
- ***Soil washing*** is a process where contaminants adsorbed "onto fine soil particles are separated from bulk soil in a water-based system on the basis of particle size. The wash water may be augmented with a basic leaching agent, surfactant, or chelating agent or by adjustment of pH to help remove organics and heavy metals. Soils and wash water are mixed ex-situ in a tank or other treatment unit. The wash water and various soil fractions are usually separated using gravity settling" [EPA 11h].
- ***Bioreactor landfills*** operate "to rapidly transform and degrade organic waste. The increase in waste degradation and stabilization is accomplished through the addition of liquid and air to enhance microbial processes. This bioreactor concept differs from the traditional 'dry tomb' municipal landfill approach. A bioreactor landfill is not just a single design and will correspond to the operational process invoked. There are three different general types of bioreactor landfill configurations:
  - *Aerobic*—Liquid leachate is removed from the bottom layer, piped to liquids storage tanks, and recirculated into the landfill in a controlled manner. Air is injected into the waste mass, using vertical or horizontal wells, to promote aerobic activity and accelerate waste stabilization.
  - *Anaerobic*—Moisture is added to the waste mass in the form of recirculated leachate and other sources to obtain optimal moisture levels. Biodegradation occurs in the absence of oxygen (anaerobically) and produces landfill gas. Landfill gas, primarily methane, can be captured to minimize greenhouse gas emissions and for energy recovery projects.
  - *Hybrid* (Aerobic-Anaerobic)—The hybrid bioreactor landfill accelerates waste degradation by employing a sequential aerobic-anaerobic treatment to rapidly degrade organics in the upper sections of the landfill and collect gas from lower sections" [EPA 11h].

Ex-Situ Thermal Treatment

This method "generally involves the destruction or removal of contaminants through exposure to high temperature in treatment cells, combustion chambers, or other means used to contain the contaminated media during the remediation process. The main advantage of ex-situ treatments is that they generally require shorter time periods, and there is more certainty about the uniformity of treatment because of the ability to screen, homogenize, and continuously mix the contaminated media; however, ex-situ processes require excavation of soils, which increases costs and engineering for equipment, permitting, and materials handling worker safety issues" [EPA 11h].

These processes use heat to separate, destroy, or immobilize contaminants. Pyrolysis and conventional incineration destroy the contaminants.

- "***Pyrolysis*** is defined as chemical decomposition induced in organic materials by heat in the absence of oxygen. Pyrolysis typically occurs under pressure and at operating temperatures above 430°C (800°F). The pyrolysis gases require further treatment. The target contaminant groups for pyrolysis are semi-volatile organic compounds (SVOCs) and pesticides. The process is applicable for the separation of organics from refinery wastes, coal tar wastes, wood-treating wastes, creosote-contaminated soils, hydrocarbon-contaminated soils, mixed (radioactive and hazardous) wastes, synthetic rubber processing wastes, and paint waste" [EPA 11h].
- ***Incineration*** is an ex-situ, heat-based technology that has been used for many years to burn and destroy contaminated materials. There are several types of incinerators discussed in Chapter 4.

Thermal desorption and hot gas decontamination are two processes that separate the contaminants from the media to which they are attached.

- "***Thermal desorption*** involves the application of heat to excavated wastes to volatilize organic contaminants and water. Typically, a carrier gas or vacuum system transports the volatilized water and organics to a treatment system, such as a thermal oxidation or recovery unit. Based on the operating temperature of the desorber, thermal desorption processes can be categorized as either high-temperature thermal desorption (320 to 560°C or 600 to 1,000°F) or low-temperature thermal desorption (90 to 320°C or 200 to 600°F)" [EPA 11h]. A parallel flow (co-current) rotary low-temperature thermal desorption system is illustrated in Figure 5.4.

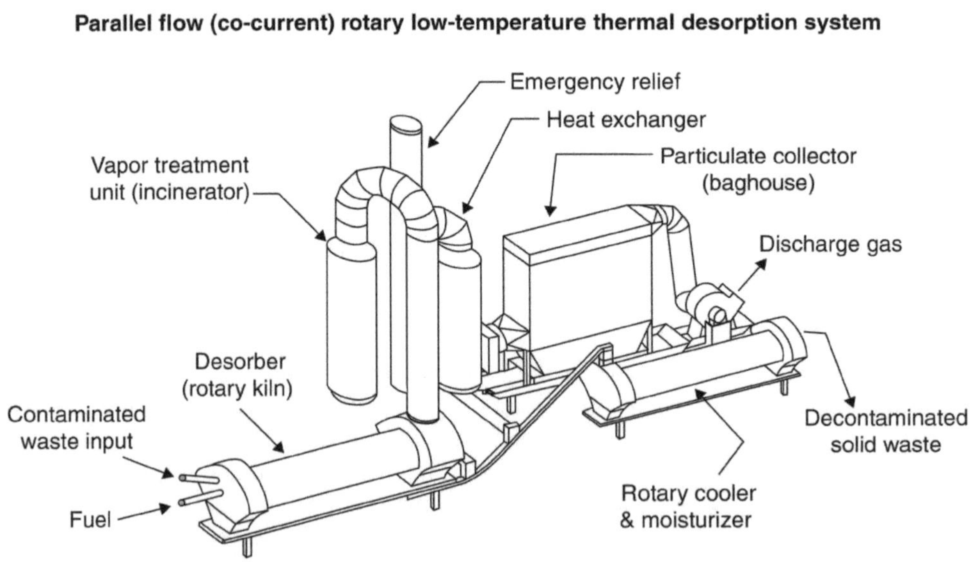

FIGURE 5.4 Thermal desorption system. (Adapted from EPA 510-B-95-007.)

- "***Hot gas decontamination*** involves raising the temperature of contaminated solid material or equipment to 260°C (500°F) for a specified period of time. The gas effluent from the material is treated in an afterburner system to destroy all volatilized contaminants. This method will permit reuse or disposal of scrap as nonhazardous material.

Other thermal technologies include:

- ***Vitrification*** technology uses an electric current to melt contaminated soil at elevated temperatures (1,600 to 2,000°C or 2,900 to 3,650°F). Upon cooling, the vitrification product is a chemically stable, leach-resistant, glass and crystalline material similar to obsidian or basalt rock. The high temperature component of the process destroys or removes organic materials. Radionuclides and most heavy metals are retained within the vitrified product. Vitrification can be conducted in-situ or ex-situ.
- ***Thermal off-gas treatment*** is one of several approaches that can be used to cleanse the off-gases generated from primary treatment technologies, such as air stripping and soil vapor extraction. In addition to the established thermal treatments, organic contaminants in gaseous form can be destroyed using innovative or emerging technologies, such as alkali bed reactors.
- ***Plasma high temperature recovery*** uses a thermal treatment process applied to solids and soils that purges contaminants as metal fumes and organic vapors. The vapors can be burned as fuel, and the metals can be recovered and recycled" [EPA 11h].

5.2 CONTAINMENT OF CONTAMINANTS IN PLACE

Although the methods mentioned below might be considered in-situ, they are really neither in-situ nor ex-situ treatment because they are designed to *confine,* rather than treat, the contaminants in their current location.

- **Capping**: Placing a cover over contaminated materials to keep them from spreading to other areas because of precipitation, wind, or physical contact by humans or animals.
- **Groundwater Cutoff/Containment Wall**: This involves digging a trench around the contaminated media to prevent the migration of contaminants from the site. The material used is impermeable, such as concrete, a clay slurry, or a plastic barrier.
- **Solidification/Stabilization:** This method is used to stabilize contaminated media that is difficult to treat by other means. Both methods are used to bind the contaminants in the soil to keep them from migrating off site, or to make them safe to dispose as landfill.

5.3 INNOVATIVE CLEAN-UP TECHNOLOGIES

The USEPA Web site at *http://www.cluin.info/products/nairt/overview.cfm* provides access to a report on innovative remediation technologies. The *Field-Scale Demonstration Projects in North America, 2nd Edition, Year 2000 Report* [EPA 00] is a revision and expansion of the 1996 USEPA publication, *Completed North American Innovative Technology Demonstration Projects*

(http://www.epa.gov/oust/cat/demorept.pdf).

The project information in the year 2000 document is now available in an online, searchable database of ongoing and completed field demonstrations of innovative remediation technologies; it is sponsored by government agencies working in partnership with private technology developers to bring new technologies into the hazardous waste remediation marketplace.

NOTE

The Field-Scale Demonstration Projects in North America report (2000) is available at http://www.clu-in.org/download/remed/nairt_2000.pdf [EPA 00].

This database contains projects through June 2000 only. Current demonstration project information is available in the separate Remediation Technology Demonstration Project Profiles database [EPA 10].

Nanotechnology in Clean-up Situations

Nanotechnology is a physical treatment that is accomplished at an incredibly small scale, and through which the materials can be manipulated to accomplish a particular purpose—in this case, the treatment of hazardous contaminants. A nanometer is one billionth of a meter—about one ten-thousandth the thickness of a human hair. Because the particles are very small compared to the contaminant particles, nanotechnology allows faster and more complete penetration of the nanoscale materials into the contaminants. Also, because the surface areas of the nanoparticles are very large when compared with their volumes, their reactivity in surface reactions can be much more effective compared with the same material at much larger sizes.

One big advantage of nanoscale materials is that they can be manipulated to create novel properties not present in particles of the same material at the micro- or macro-scale. Nanoscale materials may also exhibit altered reaction rates that cannot be explained by surface-area alone. These properties can provide enhanced contaminant contact, causing rapid reduction of contaminant concentrations.

An EPA Web site describes several success stories and an overview of the potential and current uses of nanotechnology [EPA 11i].

NOTE

Some of the clean-up technologies listed in this chapter require air pollution control technologies and/or water pollution control technologies to be applied, because some of these treatment technologies cause the release of contaminants into the air or surface of water. Chapter 6, "Air Pollution Control," describes basic air pollution technologies and their application to hazardous waste streams, and Chapter 7, "Wastewater Management," does the same for wastewater treatment technologies. Although not addressed in detail in this book, federal and state air and water pollution regulations apply to many forms of hazardous waste treatment. For further information, contact the state regulatory authority in the state where the cleanup is located.

5.4 CASE STUDIES

The following case study is an example of the problems associated with a company that containerizes and sells compressed gases.

Case Study

Compressed Gas Facility Going Out of Business

A hazardous waste inspector received a notice from the U.S. Food and Drug Administration (USFDA) that a business that supplied medical oxygen was temporarily suspended from selling containerized oxygen because of various health violations. The inspector visited the site to see if there were any waste compressed gases on site that needed to be properly disposed.

The business owner was in poor health, and although he clearly wanted to comply with the corrections mandated by the FDA, he was unable to do so, and the oxygen supply business failed. The owner had stated that medical oxygen supply was the mainstay of his compressed gas business, so his inability to keep the oxygen business caused concern about the fate of the remaining compressed gas cylinders on site. In an on-site inventory, the inspector discovered over one thousand compressed gas cylinders, many of which were visibly corroded; some were stored in trailers and other buildings on and off site.

As discussed in Chapter 2, "Identification of Hazardous Waste," waste compressed gases with pressure over 1 atmosphere, that are flammable or that will support combustion, are D001 hazardous waste. There were hundreds of oxygen and acetylene cylinders scattered throughout the property, because the facility refilled these cylinders for welders throughout the area. There were cylinders of nonhazardous gases such as nitrogen, argon, and freon, together with some fairly uncommon gases. The most poisonous gas found on site was phosgene, a gas that was used as a nerve agent in World War I, but that is still used as a reagent in the pharmaceutical industry today.

Of even greater concern were other types of compressed gases that were unstable or even explosive. One example of these is hydrogen gas, highly explosive at room temperature and pressure. Another potentially unstable and explosive gas is contained in old acetylene containers. If acetylene gas is pressurized in its pure form over 15 PSI, it breaks down or separates into carbon black and hydrogen gas, a highly unstable and explosive mixture. If acetylene gas is placed in a container with a fabric "wick" containing a solvent such as benzene or acetone in the center of the tank, it is stable up to 1500 PSI. A serious problem can arise if the acetylene cylinder is stored for a long period, because the solvent that stabilizes the gas can diffuse out of the wick and container, leaving a very unstable, explosive container.

When the business shut down a few weeks later, the facility manager vented all of the dangerous containers on site (except for the poisonous gases). Those empty containers were hauled off site for scrap metal recovery. This incident had a satisfactory ending in that all of the gas cylinders were removed without incident, the building was demolished, and a new business has been built on the same site.

The hazardous wastes generated in this case were predominantly D001 (ignitable), but because they were released directly to the environment and they had no measurable weight, after the containers were vented, it was as though the wastes had never been generated.

The next case study concerns a company established to recover mercury from hearing aid batteries, switches thermostats, and other mercury-containing equipment, and to sell the recovered mercury for future manufacturing.

Case Study

Mercury Recovery Facility Creates Federal Clean-up Site

Beginning in 1955, the Mercury Refining Company (MERECO) in upstate New York accepted mercury-containing wastes from batteries, thermometers, dental amalgam, etc., and refined the mercury for resale.
The mercury was reclaimed using retort (condensation) furnaces at the facility. This facility was a commercial hazardous waste facility because hazardous waste was received from off site for on-site storage.

Before 1980, waste contaminated with mercury was dumped over an embankment of an unnamed tributary to a creek. Contaminated groundwater ran off the site. In 1981, there was a structure fire at the facility, and a large amount of water used for fighting the fire ran off the site. The NYSDEC sampled the area around the tributary and found polychlorinated biphenyls (PCBs) and mercury contamination on the southern edge of the property and the embankment of the tributary. In 1983, the USEPA placed the MERECO site on the National Priorities List of the most contaminated sites in the country.

In 1998, MERECO discontinued reclaiming mercury, but continued reclaiming precious metals at the facility.

After several interim cleanup and containment actions, in 1999, at the request of the NYSDEC, the USEPA took over as lead agency and initiated and completed a remedial investigation and feasibility study of the site. The final costs for the investigation and cleanup at the site are approximately $11,500,000. Even after the cleanup, residual mercury pollution will remain downstream of MERECO in the Patroon Creek [EPA 11j].

The next case study is about a little-known chemical that becomes highly unstable and explosive when it is stored for a long time.

Case Study

Picric Acid Found in an Office Building

Picric acid [$C_6H_2(NO_2)_3OH$], also known as trinitrophenol, is an unstable acid used primarily as a chemical reagent and as a booster to detonate other, less sensitive explosives, such as TNT (trinitrotoluene). Other uses of picric acid have been as an antiseptic, a yellow dye, and in the synthesis of a powerful insecticide. Picric acid was used in bombs and grenades in World War I.

The salts of picric acid can be shock and friction sensitive when this substance becomes crystallized. When a bottle of picric acid is stored for a long time and the acid crystallizes under the cap, an explosion can result from someone simply unscrewing the cover.

In one incident, a biological fisheries lab had used the yellow dye property of picric acid to mark the sides of fish during a study. The unused acid was then stored in a one gallon jug in the basement of an office building for several years. When the old bottle was discovered, it was in a shock-sensitive box, supported in all directions by springs. Thankfully, the container was clearly labeled, and the workers inquired about the safety issues before disturbing the container. After inquiring and finding out about the acid's explosive and unstable nature, a specialty waste management company was called in to safely remove the container.

This incident ended well; the container was safely removed from the building and was detonated in a safe location. Had the bottle detonated inside the building, the explosive force would have likely caused considerable damage.

The hazardous waste generated in this case is D002 (see Table 2.1 in Chapter 2), because picric acid is corrosive, but should also be listed as a (D001) flammable solid.

The next study is about a disturbing nationwide trend: the manufacture of methamphetamines (crystal meth) in residences and remote areas.

Case Study

Methamphetamine Laboratories

Hazardous waste inspectors are sometimes asked to visit abandoned methamphetamine (meth) labs to help evaluate the hazardous wastes left behind and to give advice on cleanup. There is a grave danger of explosions and fires occurring at unattended or abandoned meth labs for at least two reasons:

- Sometimes explosive traps are set by perpetrators running an active lab to prevent intruders from entering the lab.
- Large amounts of solvent (alcohol and acetone) are used to strip the pseudoephedrine/ephedrine for the formulation of the methamphetamine. Many meth lab explosions are the result of different solvents evaporating, in a poorly ventilated area, and being exposed to a spark or an open flame.

FIGURE 5.5 Chemicals from a methamphetamine lab. (Adapted from EPA-560-F-05-232.)

The pollution resulting from a meth lab can be very detrimental to the environment, in that when these operations are abandoned, the perpetrators leave behind containers of solvents and other wastes that are often dumped by the owners or otherwise eventually leak into the environment.

The hazardous wastes left behind are commonly ignitable (D001) (see Table 2.1) and F-list (see Table 2.2) waste, depending on the exact solvent(s) left behind.

The next case study is about a long-abandoned fireworks factory discovered in a wooded area. The cleanup had many tense moments.

Case Study

Abandoned Fireworks Factory

A hazardous waste inspector received an anonymous call that an abandoned fireworks factory had been discovered. When arriving at the site, the inspector discovered a very old concrete building that had once been the fireworks factory, filled with unmarked containers. Several old, collapsing sheds, cars, and trucks were scattered around the five-acre property, and the sheds and each vehicle was packed full of old, long-expired fireworks. The abandoned factory was also full of old corroded containers of black gunpowder, wet fireworks, and the materials used to make fireworks.

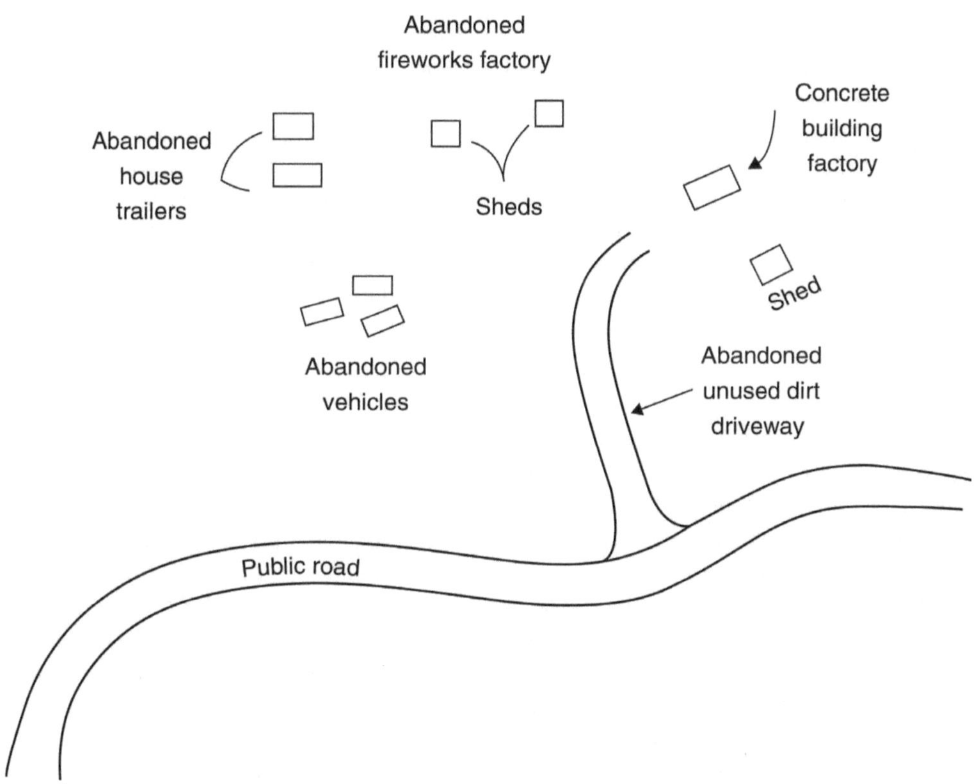

FIGURE 5.6 Abandoned fireworks factory. (Drawing by author.)

The following questions should have been asked for this site:

Q: What is the inspector's highest priority?
A: Safety. Abandoned fireworks are flammable, explosive, and if old or wet, they can be unstable.
Q: What types of hazardous wastes might be found on site?
A: D001 (Flammable), D003 (Reactive), D004-38 (metals) and some F wastes (solvents)

Q: What should the inspector do first?
A: Stay clear of all fireworks and call the office to tell them what had been found.
Q: What should be done next?
A: Call police or fire department for help in securing the site.
Q: Which government agencies would be ultimately responsible for the cleanup?
A: Federal Emergency Management Agency (FEMA), state emergency agency, local police and fire departments, Federal Bureau of Alcohol, Tobacco and Firearms (ATF), Military Demolition Squad, federal, state and local environmental and health agencies.

This situation was real, and the outlined steps were taken. Thankfully, no one was injured during the investigation, removal, transportation, and destruction of the fireworks. After the site was discovered and the environmental agency was contacted, the local fire department and police were called. The federal and state emergency management agencies implemented an emergency management plan, asking for assistance from FEMA, ATF, U.S. Army, USEPA and the state environmental agency.

How was it disposed? Under RCRA, these explosive wastes should have been properly packaged and transported to a licensed hazardous waste TSDF and burned.

However, given the unstable nature of the waste, and the hazards associated with packaging the wastes, the ATF supervised while the U.S. Army Munitions Unit loaded several dump truck loads of fireworks and materials and transported and burned them in a gravel pit. This case ended without incident, other than the expense to the state for the emergency management measures. Needless to say, it could have ended badly if the abandoned factory had been vandalized, especially by fire.

In this chapter, you learned about treatment technologies for contaminated hazardous waste sites, containment technologies for contaminated sites, and a few innovative technologies for contaminated sites. You also read case studies about a compressed gas facility going out of business, a poorly operated mercury recovery facility creating a cleanup site, picric acid in an office building, the dangers of methamphetamine labs, and an abandoned fireworks factory.

In the next chapter, you will learn the history of air pollution in the United States, learn about the structure of the atmosphere, study sources of air pollutants, learn classification of air pollutants, study the effects of air pollution, learn air pollution control technologies, read about performance standards for incinerators, study automobile pollution and control technologies, and discover the relationship between air pollution control and hazardous waste. You will also read case studies on complaints by neighbors causing more hazardous waste and air pollution, dry cleaners driven by regulations, and trial burns for hazardous waste combustors.

EXERCISES

1. Why do we need specialized remediation or cleanup technologies?
2. What is the difference between in-situ and ex-situ treatment?
3. What is the difference between air stripping and air sparging?
4. What are the requirements for bioremediation to work effectively?

5. Where is soil vapor extraction (SVE) most effective?
6. What conditions are adverse to phytoremediation?
7. Name a few examples of solid-phase bioremediation.
8. What is the main advantage of all bioreactor landfills?
9. What is pyrolysis and how is it used in waste treatment?
    a. What is one drawback of this technology?
10. Name at least two situations where a containment technology would be necessary at a contaminated hazardous waste site.

REFERENCES

[EPA 00] USEPA Office of Solid Waste and Emergency Response EPA 542-B-00-004 June 2000 (5102G), online at *http://www.clu-in.org/download/remed/nairt_2000.pdf*, (accessed May 20, 2011).

[EPA 10] Remediation Technology Demonstration Project Profiles database, last updated June, 2010, online at *http://www.clu-in.org/products/demos/search/vendor_search.cfm*, (accessed May, 20 2011). [EPA 11h] USEPA Superfund Remediation Technologies, available online at *http://www.epa.gov/superfund/remedytech/remed.htm*, (accessed May, 20 2011).

[EPA 11i] USEPA Technology Innovation and Field Services Division, Nanotechnology Applications for Environmental Remediation, online at *http://www.clu-in.org/techfocus/default.focus/sec/Nanotechnology%3A_Applications_for_Environmental_Remediation/cat/Overview/*, (accessed May 20, 2011).

[EPA 11j] USEPA Region 2 Fact Superfund Facts on Mercury Refining, online at *http://www.epa.gov/region2/waste/fsmercur.htm*, (accessed May 20, 2011).

CHAPTER 6

Air Pollution Control

In This Chapter

- The history of air pollution in the United States
- The structure of the atmosphere
- Sources of air pollutants
- Classification of air pollutants
- Effects of air pollution
- Air pollution control technologies
- Performance standards for incinerators
- Automobile pollution and control technologies
- The relationship between air pollution control and hazardous waste
- Case studies

Hazardous waste regulations apply to the storage, transportation, and treatment of hazardous waste; in general terms, however, these regulations do not apply when the waste is discharged to an air pollution control system or a wastewater treatment plant. That does not mean the waste ceases to be hazardous when discharged to another system—it means the discharge has crossed a "regulatory boundary" and is no longer regulated by the RCRA program. In the United States, industrial air pollution discharges are regulated under the Clean Air Act and wastewater discharges are regulated under the Clean Water Act.

This chapter introduces air pollution control as it pertains to hazardous waste.

Numerous portions of this chapter have been quoted directly with permission from Dr. S. Amal Raj , Introduction to Environmental Science and Technology. Laxmi Publications Pvt. Ltd. 2008. 27–41.

6.1 HISTORY OF AIR POLLUTION IN THE UNITED STATES

"Air pollution initially was recognized more as a nuisance than as a threat to human health. Some laws, however, were enacted to prevent air pollution as early as 1306. In that year, King Edward the 1st of England ordered that the burning of sea coal in craftsman's furnaces be prohibited because of the foul-smelling fumes produced. Centuries later, Queen Elizabeth the 1st banned the burning of coal in London whenever Parliament was in session.

As the years passed, air pollution got worse, and yet it was still not widely recognized as a threat to human health. Although there were some scientists and health professionals who recognized air pollution as a public health problem, most of the early control efforts were targeted at the aesthetic or welfare effects of air pollution. In the late 1800s and early 1900s, many smoke control ordinances were enacted in England and the United States. These laws were some of the first uniform statutes enacted for the control of air pollution. Our modern air pollution control program can be said to have evolved from these early ordinances.

The delay in recognizing air pollution as a health risk was partly a result of the nature of air pollution. Air pollution is usually not recognizable as is water pollution; therefore, it can be ignored as a health threat until the problem reaches crisis proportions. Air pollution episodes from the buildup of pollutants in the air influenced the development of air pollution programs in the United States.

In the 1940s, air pollution received greater attention in the United States when smog was noticed in Los Angeles. Visibility was only three blocks and people suffered from smarting eyes, respiratory discomfort, nausea, and vomiting. California passed the first state air pollution law in 1947, and the first National Air Pollution Symposium in the United States was held in 1949. Initially, municipal governments were responsible for the passage and enforcement of such legislation.

The federal government of the United States began efforts to control air pollution in the 1950s with the passage of the Air Pollution Control Act of 1955. This was the first federal air pollution law, and it mandated federal research programs to investigate the health and welfare effects of air pollution. It also authorized the federal government to provide technical assistance to state government. Additional legislation was passed in 1963 with the purpose of empowering the Secretary of Health, Education, and Welfare to define air quality criteria based on scientific studies and provided grants to state and local air pollution control agencies. This Clean Air Act (CAA) replaced the Air Pollution Control Act of 1955. In 1965 the Motor Vehicle Air Pollution Control Act was directed to establish auto emission standards. In 1967 the Federal Air Quality Act was enacted and established framework for defining 'air quality control regions' based on meteorological and topographical factors of air pollution.

In 1970 President Nixon created the Environmental Protection Agency (EPA) by Executive Order. An executive order is an order issued by a government's chief executive, intended to give attention to a certain law or body of laws and directs federal agencies how to implement them. The formation of EPA marked a dramatic change in national policy regarding the control of air pollution. Whereas previous federal involvement had been mostly in advisory and educational roles, the new EPA emphasized stringent enforcement of air pollution laws. The EPA was assigned the daunting task of repairing the damage already done to the natural environment and establishing new criteria to guide Americans in making a cleaner environment a reality. A few weeks later the United States Congress passed the Clean Air Act Amendments (CAAA) of 1970. The passage of the CAAA of 1970 marked the beginning of modern efforts to control air pollution.

The Clean Air Act (CAA) is the law that defined EPA's responsibilities for protecting and improving the nation's air quality. Congress drafted the Clean Air Act in 1963, establishing funding for the study and cleanup of air pollution. However there was no comprehensive federal response to address air pollution until Congress passed a much stronger Clean Air Act in 1970. In the same year, Congress created the EPA and gave it the primary role in carrying out the law" [EPA 11k].

An electronic version of the Clean Air Act can be found at

http://www.epa.gov/oar/caa/.

A layman's-language guide to the Clean Air Act is also available and can be found at

http://www.epa.gov/air/caa/peg/.

6.2 STRUCTURE OF THE ATMOSPHERE

Definition: "In general, air pollution may be defined as the presence of substances in such a concentration that makes the air harmful or dangerous to breathe or to cause damage to plants, animals and the environment" [Raj 08].

The air we breathe, like the water we drink, is necessary for life. As with water, we have to be assured that breathing the air around us will not cause us harm. We expect to breathe "clean air." Pure air in the atmosphere is a mixture of gases, which contains by volume: 78.1% nitrogen (N_2), 20.9% oxygen (O_2), 0.93% argon, 0.032% Carbon Dioxide (CO_2), 0.002% neon, and trace amounts of other gases, most of which are inert. But such air is not found in nature, and is of interest only as a reference, as pure water (H_2O) would be.

If this is pure air, then it may be useful to define as pollutants those materials (gases, liquids, or solids) that, when added to pure air at sufficiently high concentration, will cause adverse effects. The pollutant emitted into the atmosphere must travel through the atmosphere to reach people, animals, plants, or materials to have an ill effect. In air pollution, wind (air currents) is the means of transport of pollutants, whereas, in water pollution, pollutants are carried by water currents.

"The earth's atmosphere can be divided into easily recognizable strata, depending on the temperature profile. Figure 6.1 shows a typical temperature profile for four major layers. The layer of greatest interest in pollution control is the troposphere, since this is the layer in which most living things exist. The temperature here usually decreases with altitude. More than 80% of the air is within this well-mixed layer. On top of the troposphere is a layer of air where the temperature profile is inverted, and in the stratosphere little mixing takes place. Pollutants that migrate up to the stratosphere can stay there for many years. The stratosphere has a high ozone (O_3) concentration, and ozone absorbs the sun's short-wave ultra-violet (UV) radiation. Above the stratosphere are two more layers, the mesosphere and the thermosphere, which contain only about 0.1% of the air. Other than the problems of global warming and stratospheric ozone depletion, pollution problems occur in the troposphere" [Raj 08].

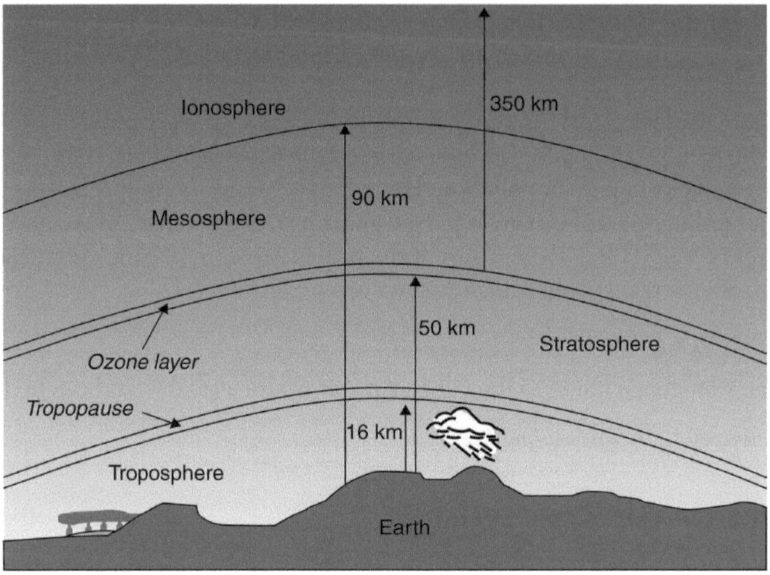

FIGURE 6.1 Structure of the atmosphere. (Adapted from EPA Appendix G - Atmospheric and Meteorological Concepts.)

6.3 SOURCES OF AIR POLLUTANTS

Air pollution can be natural or might be the result of various human or animal activities such as domestic, industrial, and agricultural.

Natural Sources

Air contains natural contaminants such as: pollen grains, fungal spores, salt spray caused by wind on waves, bacteria, smoke, and dust particles from forest fires, volcanic eruptions, wind-blown dust, naturally occurring carbon monoxide (CO) from the breakdown of methane, and hydrogen sulfide (H_2S) and methane from the anaerobic degradation of organic matter. (For example, organic emissions from trees and animal feces.)

Anthropogenic (Man-Made) Air Pollution Sources

"In contrast to the natural sources of air pollution, there are contaminants of anthropogenic origin. The sources of these pollutants are numerous and are classified as follows:

Mobile Sources Include:

- Line sources, such as: highway vehicles, aircraft, railroads, and ships
- Area sources such as airports, rail yards, and harbors

Stationary Point and Area Combustion Sources Include:

- Point sources like: thermal power plants; and fuel combustion of oil, natural gas, and coal
- Area sources such as: residential heating, institutional heating, commercial heating, gas stations, consumer products, and farms

Industrial Point Sources Include:

- Chemical, metallurgical, pulp and paper, petroleum refineries, cement plants, auto body shops, and dry cleaners

Solid Waste Disposal Such As:

- On-site incineration, off-site incineration, and landfill off-gassing" [Raj 08]

TABLE 6.1 Sources of industrial air pollutants

Sources	Types of Pollutants
Industrial combustion	Smoke, fly ash, SOx, NOx, CO
Chemical process, paper mills, cement plants, fertilizer plants, etc.	SOx, NOx, NH_3, CO, Volatile Organic Compounds (VOCs)
Petroleum refineries	SO_2, H_2S, NH_3, CO, hydrocarbons, mercaptains
Mineral processing	Dust, fumes, SO_2, CO, fluorides, and organic vapors

Source: Permission granted courtesy of Laxmi Publications Pvt. Ltd.

6.4 CLASSIFICATION OF POLLUTANTS

"Air pollution can be classified according to the origin, chemical composition, and state of matter. A breakdown of each of these origins is as follows:

Origin

According to their origin, pollutants may be considered as primary or secondary.

- ***Primary air pollutants*** such as oxides of sulfur (SOx), oxides of nitrogen (NOx), and Volatile Organic Compounds (VOCs) are those directly emitted into the atmosphere and found there in the same form.
- ***Secondary air pollutants*** are those which are formed in the atmosphere by the interaction of two or more primary pollutants, by processes such as photochemical reaction, hydrolysis and oxidation. Examples of secondary air pollutants are: ozone, peroxy acetyl nitrate (PAN), photochemical smog, and acid mist.

Chemical Composition

Pollutants can be further classified according to their chemical composition, as either organic or inorganic

- ***Organic compounds***, containing carbon, hydrogen, and other elements such as oxygen, nitrogen, phosphorous, and sulfur. Hydrocarbons are organic compounds containing only carbon and hydrogen.
- ***Inorganic compounds*** found in the atmosphere include: carbon monoxide (CO), carbon dioxide (CO_2), sulfur oxides (SOx), nitrogen oxides (NOx), ozone, hydrogen fluoride, hydrogen chloride, and metal (mercury and oxides, lead, nickel, etc.).

State of Matter

Pollutants can be classified on the basis of the state of matter as particulates and gaseous.

Particulate Pollutants

Particulate pollutants include: dust, fumes, smokes, fly ash, mist, spray, and smog.

Particulates may be further classified according to their physical, chemical, and biological characteristics.

- **Physical properties**: particle size, mode of formation, and settling behavior,
- **Chemical properties**: organic, and inorganic,
- **Biological properties**: survival time" [Raj 08]

NOTE *Under quiescent conditions, particulate pollutants will settle out of the atmosphere due to gravitational force.*

Physical Properties

"Mode of formation—Particles can be classified according to their mode of formation as dust, smoke, fumes, fly ash, mist, and spray. The first four are solids and the last two are liquid particles.

Dust is solid particles of natural or industrial origin, usually formed by the disintegration process. Dust may also come directly from the processing or handling of materials, such as coal, cement, or grains. Dust particles range in size from a lower limit of 1 μm up to about 200 μm and larger.

Smoke consists of fine solid particles formed due to incomplete combustion of organic matter such as wood, coal, etc. The particle size varies from 0.5 to 1 μm.

Fly ash consists of finely divided non-combustible particles contained in flue gases arising primarily from combustion of coal. Fly ash can also be generated from the incineration of municipal garbage or other fuels. The particle size varies from 1 to 100 μm.

Fumes are solid particles generated by the condensation of vapors by sublimation, distillation, calcinations, or other chemical reactions. Examples of fumes are zinc oxide and lead oxide resulting from the condensation and oxidation of metals volatilized under high temperatures. They range in size from 0.03 to 0.3 μm.

Mist consists of liquid particles which may arise from the condensation of a vapor. Mist is usually less than 10 μm.

Spray consists of liquid particles generated by atomization of parent liquids such as pesticides and herbicides. The size of the particles usually ranges between 10 and 1,000 μm.

Settling Characteristics

On the basis of settling characteristics, particles can generally be classified as suspended and settleable. Suspended particulate matter can be designated symbolically as SPM, for particulate matter less than 10 μm only. They remain suspended in the atmosphere for a long period of time.

Chemical Characteristics

Atmospheric particles contain both organic and inorganic components. Some of the more common organic particulates include organic acids, phenols, and alcohols. Common inorganic particulates include metals such as lead (Pb), manganese (Mn), zinc (Zn) and Radium.

Biological Characteristics

The biological particulates in the atmosphere include protozoa, bacteria, fungi, spores, pollens, and viruses. Microorganisms generally survive only for a short period in the atmosphere because of a lack of nutrients and UV radiation from the sun. However, certain bacteria and fungi form spores and can survive for a long period of time.

Gaseous Pollutants

Gaseous pollutants are formless fluids that completely occupy the same space into which they are released. They behave much like the surrounding air, and do not settle out of the atmosphere. Gaseous pollutants include: carbon monoxide (CO), carbon dioxide (CO_2), sulfur oxides (SOx), nitrogen oxides (NOx), hydrogen chloride (HCl), and oxidants (VOCs)" [Raj 08].

6.5 EFFECTS OF AIR POLLUTION

"Air pollution causes aesthetic loss, economic loss, personal discomfort, and health hazard.

Aesthetic Effects

Aesthetic effects include loss of clarity in the atmosphere as well as the presence of obnoxious odors. Atmospheric clarity loss may be caused by particulates and smog. Obnoxious odors encompass a range of potential air pollutants in gaseous form (hydrogen sulfide, mercaptans, etc.).

Economic Losses

Economic losses resulting from air pollution are:

Soiling

Soiling represents the general dirtiness of the environment that requires more frequent cleaning. Examples include more frequent washing of clothes, washing of vehicles, and repairing of structures. Soiling is due to particulate matter being deposited with the key component being settleable particulate matter.

Damage to Vegetation

Most of the damage to vegetation is due to excessive exposure of gaseous air pollutants. It has been reported that settleable particulates also disrupt normal functional processes within vegetation and thus undesirable effects take place. An example is the deposit of settleable particulate around cement plants. The air pollutants of greatest concern to agriculture are ozone (O_3), sulfur dioxide (SO_2), nitrogen dioxide (NO_2) peroxy acetyl nitrate (PAN), etc. Plants are damaged by both acute injury, in which pollutants attack cells, and chronic effects related to disruption of chlorophyll synthesis. The low concentration of pollutants for a long period of exposure may significantly decrease crop yields without visible damage.

Deterioration of Exposed Materials

The deterioration of exposed materials includes the corrosives of metals, weathering of building materials, discoloration of paint, cracking of rubber weakening of fabrics, and fading of dyes. Sulfur dioxide accelerates the corrosion of metals, necessitating more frequent repainting of metal structures. The weathering of building materials is attributed to the effects of acidic mists formed in the stratosphere as a result of oxidation processes combined with water vapor. Examples of acidic mists include sulfuric acid and nitric acid.

Health Effects

The health effects of air pollution range from personal discomfort to actual health hazard. Personal discomfort is characterized by eye irritation and irritation to individuals with respiratory difficulties. Eye irritation is associated with oxidants such as ozone, PAN and others. People who have respiratory diseases such as asthma, bronchitis, and sinusitis suffer increased discomfort due to oxidants, nitrogen oxides, and particulate matter" [Raj 08].

The effects of various air pollutants are summarized in Table 6.2.

TABLE 6.2 Effects of air pollutants

Type of Pollutant	Effects
Particulate matter	Obscure vision, aggravates respiratory diseases, corrodes metals
Sulfur oxides	Irritates respiratory tract, corrodes metals, discolors paints, weakens fabrics, leather, and paper
Nitrogen oxides	Irritates eyes and nose, stunts plant growth, causes leaf damage
Carbon monoxide	Causes headaches, giddiness, and nausea, absorbs into blood, reduces oxygen transport capability
Oxidants	
(1) Ozone	Disturbs lung function, irritates eyes, nose, and throat, discolors leaves, damages and fades textiles, accelerates cracking of rubber
(2) Peroxy acetyl nitrate (PAN)	Irritates eyes, causes shortness of breath and headaches, impairs pulmonary function, and discolors lower leaf surfaces

Source: Permission granted courtesy of Laxmi Publications Pvt. Ltd.

6.6 AIR POLLUTION CONTROL TECHNOLOGIES

"Basically, three means are available to control pollutant discharges into the atmosphere and to reduce their detrimental effects. They are:

- ***Utilizing control equipment***, which removes pollutants before they are released into the atmosphere;
- ***Source reduction,*** which can be achieved by raw material changes, operational changes, or modification or replacement of process equipment; and
- ***Dilution of pollutants*** at the source by providing tall stacks. A taller stack does not dilute pollutants. It allows more time for dilution before the pollutants reach the ground.

Control of Particulate Matter

Air pollution control technology for particulate matter from stationary sources includes the following:

- Gravity settlers
- Cyclones
- Wet collectors or scrubbers
- Electrostatic precipitators (ESP)
- Fabric filters

The following information provides brief details on the most common types of control devices for removing particulate matter.

Gravity Settlers

A gravity settler removes particulate matter by slowing down the velocity of the discharge so that the solid particulates will drop out of the gas stream (Figure 6.2). The gravity settler is a simple and inexpensive device, often used as a precleaner to remove coarse particles, in conjunction with other air pollution control devices. Typically gravity settlers are more effective on particle sizes greater than 100 µm" [Raj 08].

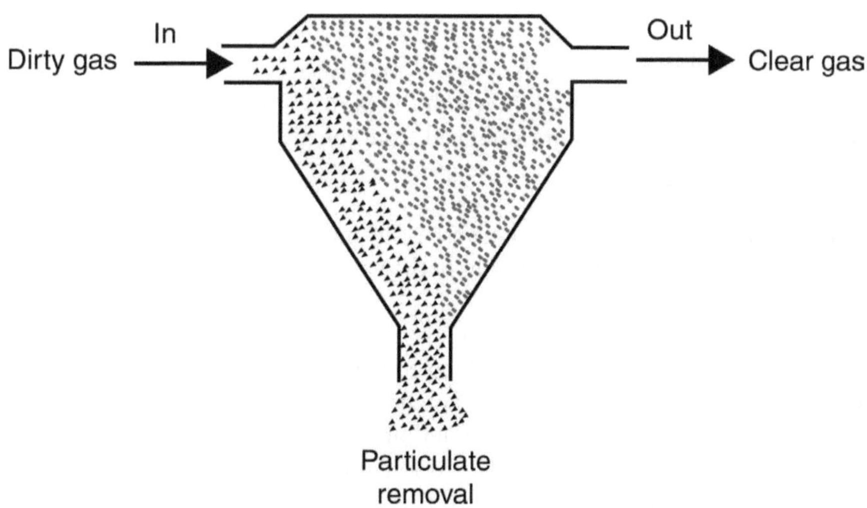

FIGURE 6.2 Schematic representation of a gravity settler. (Permission granted courtesy of Laxmi Publications Pvt. Ltd.)

Cyclones

Cyclone collectors employ centrifugal force instead of gravity to separate particles from the gas stream. Because centrifugal forces can be generated that are several times greater than the gravitational force, particles that are much smaller can be removed by this technology.

A cyclone consists of a cylindrical shell with tangential gas entry and an inverted cone attached to the base (Figure 6.3). The tangential entry imparts a whirling motion which causes the particles to be thrown toward the wall, where they collect and slide down toward the conical collector. The collection efficiency of a cyclone depends on the magnitude of centrifugal force exerted on the particles.

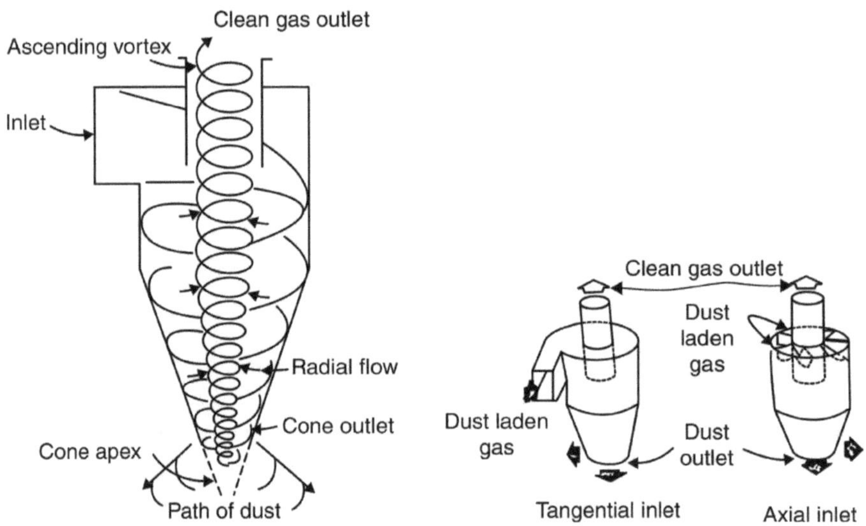

FIGURE 6.3 Typical cyclone collector. (Adapted from EPA-530-R-94-014.)

Wet Collectors or Scrubbers

"A scrubber is a collection device in which the particles are washed out of the gas flow by a water spray. In general, particles are removed from the gas stream by one or a combination of the following mechanisms:

- **Inertial impaction**: A particle, because of its inertia, may be unable to adjust to the rapidly changing curvature of streamlines in the vicinity of a droplet and may cross the streamline to hit the droplet.
- **Direct interception**: The gas flow path in which the particle is flowing may pass sufficiently close to the water droplet making contact with the surface and be collected.
- **Diffusion**: Sub-micron particles do not flow in streamlines, but wobble randomly, due to collisions with molecules of gas. Random Brownian motion causes incidental contact between the particle and the droplet. At low flow and velocity, particles spend more time near droplet surfaces, thus enhancing diffusional collection.

NOTE *Brownian motion or movement is the random motion of small particles in a fluid, either liquid or gas.*

- **Electrostatic attraction**: Particles having opposite charges are attracted to each other and are captured.
- **Gravitational force**: Some of the larger particles are removed by gravitational pull. Gravity is important for larger particles at low flow velocities.
  - Types of scrubbers
    - Spray towers
    - Packed bed towers
    - Venturi-scrubbers
  - Spray Towers [shown in Figure 6.4] are low-cost scrubbers that can be used to remove both gaseous and particulate contaminants. The spray tower is a vertical column in which the gas passes upward and the liquid solvent passes downward. Liquid droplets are formed by the process of atomization. As the gas flows upward, entrained particles collide with liquid droplets sprayed across the flow passage. By the time the gas reaches the top of the column, most of the gaseous contaminants dissolve in the solvent and the particulates are collected on the liquid droplets. The contaminant-loaded solvent settles by gravity at the bottom of the chamber, where it is discharged to a wastewater treatment system [see Chapter 7, "Wastewater Management"]. In general, the smaller the droplet size and the greater the turbulence, the more chance for absorption of the gas. Production of fine droplets requires the use of high-pressure spray nozzles. If water is the solvent, the application is limited to a few organic gases, such as ammonia (NH_3), chlorine (Cl_2), and sulfur dioxide (SO_2).

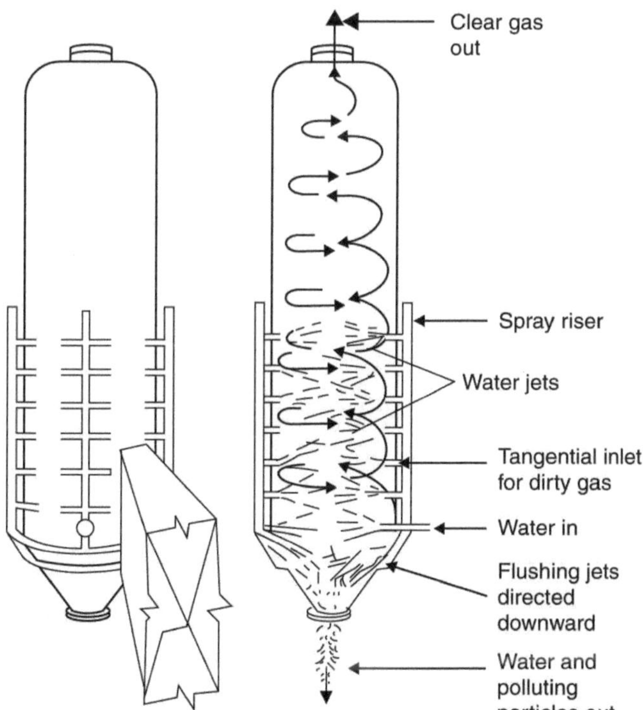

FIGURE 6.4 Spray tower. (Permission granted courtesy of Laxmi Publications Pvt. Ltd.)

- Packed bed towers are also known as Packed Bed Scrubbers, or Packed Towers. They are vertical cylindrical towers in which the contaminated air stream comes into contact with a solvent that absorbs or chemically reacts with the pollutants. The air is cleaned and then discharged to the atmosphere, with the contaminated liquid treated in a wastewater treatment system or chemically treated or recycled to recover useful chemicals, which can be used or reused.
- Venturi scrubbers generally consist of three separate sections: a narrow throat section where the liquids enter, a second area where the liquids and gases converge, and a section where they diverge. The inlet gas stream enters the converging section and, as the area decreases, gas velocity increases. Liquid is introduced either at the throat or at the entrance to the converging section.
- The Venturi principle is employed as the inlet gas moves at extremely high velocities in the narrow throat section and shears the liquid from its walls. This produces a huge number of tiny droplets (like fog). When the inlet gas stream mixes with the fog of tiny liquid droplets, the particles and gases are removed in the throat section. When the inlet stream exits the diverging section, it is forced to slow down" [Raj 08].

Electrostatic Precipitators (ESP)

"Electrostatic precipitation uses the force of an electric field on electrically charged particles to segregate particulate matter from a waste gas stream. The particles are deliberately charged and passed through an electric field causing the particles to migrate toward an oppositely charged electrode, which acts as a collection surface. The accumulated deposits are removed from the collection surfaces from time to time. ESPS are of two types: plate type; and pipe type.

- ***Plate type ESPs*** consists of parallel vertical grounded steel plates collecting electrodes together with an array of discharge wires mounted on a plane midway between the plates (see Figure 6.5). The dirty gas with particulates passes between the plates. As the gas flows between the parallel plates, electrostatic forces cause the particulates to migrate to the collector electrode where they stick. The clean gas then emerges on the other side. The collected particulates are removed from the collector electrode by either water washing or rapping it periodically" [Raj 08].

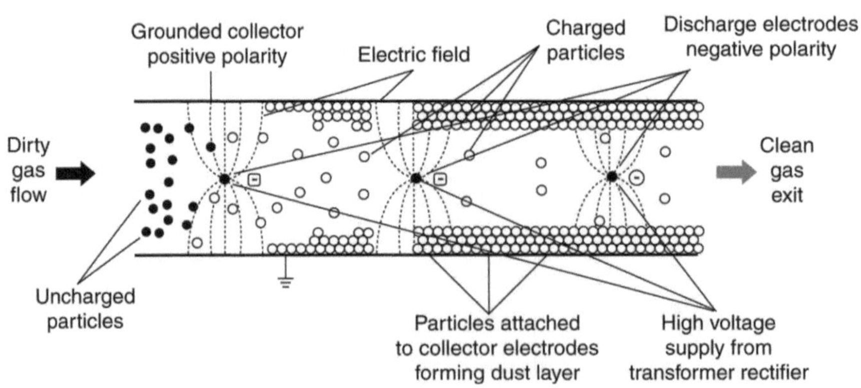

FIGURE 6.5 Schematic representation of plate type electrostatic precipitator. (Adapted from EPA-600/R-04/072.)

- ***Pipe type ESPs***, are also called wire-pipe ESPs or tubular ESPs. The exhaust gas flows vertically through conductive tubes, generally with many tubes operating in parallel. Similar to the plate type ESP, the particles are attracted to the collector electrode, and the clean gas emerges on the other side.

ESPs are characterized by high collection efficiencies, even for sub-micron particles. They can handle large gas volumes with low pressure drops.

Fabric Filters

"A fabric filter is a membrane (cloth, wire mesh, etc.) with holes smaller than the dimensions of the particles to be retained. As the fine particles are caught on the sides of the holes of the filter, they tend to close the holes and make them smaller. As the amount of collected particles increases, the cake of collected materials becomes the filter and the filter medium (fabric) now supports the cake.

When a dirty gas stream flows through the filter medium, the solid particulates deposit on the face of the medium. The cleaned gas flows through both the filter cake and the medium, and the pressure drops, causing the gas to expand.

As the most commonly used surface filters have an enclosing sheet metal structure in the shape of a house, it is also called a baghouse. It consists of a number of cylindrical cloth bags that are closed at the top. They hang from a support.

The bag is made of natural or synthetic fibers. Synthetic fibers are widely used for filtration because of their low cost, better temperature and chemical resistance characteristics, and small fiber diameter. Bag life varies between one and five years. The hopper at the bottom serves as a collector for dust" [Raj 08]. (See Figure 6.6.)

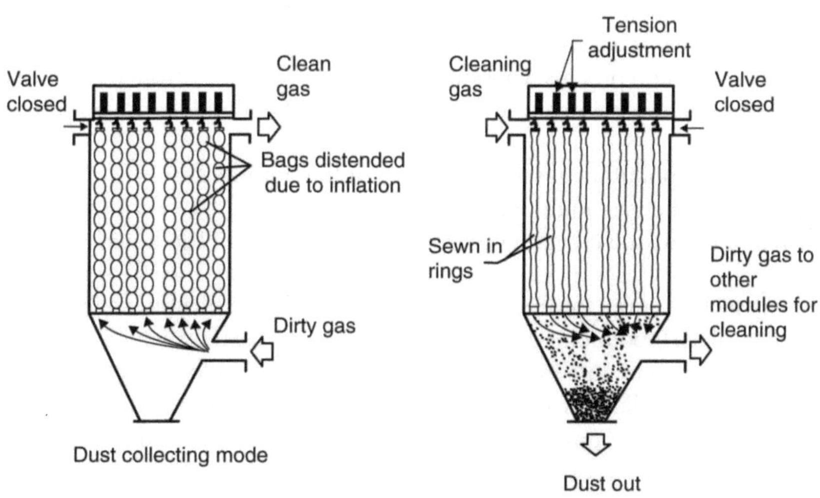

FIGURE 6.6 Reverse-air baghouse. (Adapted from EPA-530-R-94-014.)

"The gas enters through the inlet chamber and larger particles fall into the hopper because of gravity. The gas then flows upward into the filter bags and then outward through the fabric. leaving the particulate matter as a cake on the inside of the bags. A dust cake builds on the inside bag surface during filtration. Efficiency during cake formation (pre-coat) is low, but increases as the cake is formed, until a final efficiency of around 99% is obtained.

Filter cleaning The accumulation of particulates increases the air resistance of the filter and therefore, the filter bags need to be cleaned periodically. They can be cleaned by rapping, shaking, or vibration, or by reverse air flow, causing the filter cake to be loosened and to fall into the hopper below.

Filter medium While selecting a filter medium for baghouses, the following properties are taken into consideration: gas temperature, gas composition, gas flow rates, size of the particulate matter, and concentration of particulate matter.

The commercially available fabric materials are cotton, wool, dacron, wool, nylon, polypropylene, fiber glass, etc." [Raj 08].

Control of Gaseous Contaminants

"Major air pollutants are gases such as carbon monoxide (CO), nitrogen oxides (NOx), sulfur oxides (SOx), and volatile organic compounds (VOCs). In general, the concentrations of pollutants in the waste air stream are relatively low, but the emissions can still exceed the regulatory limits. Removal of gaseous pollutants can be achieved by the following methods:

- Absorption
- Adsorption
- Condensation
- Biofiltration
- Combustion

Absorption

Absorption, also called scrubbing, involves transferring pollutants from a gas phase to a contacting solvent. This is a mass transfer phenomenon in which the gas dissolves in the liquid. Mass transfer is a diffusion process wherein the pollutant gas moves from the points of higher concentration to points of lower concentration. The removal of pollutant gas takes place in three steps:

- Diffusion of the pollutant gas to the surface of the liquid
- Transfer across the gas-liquid interface
- Diffusion of the dissolved gas away from the interface into the liquid

Equipment such as spray towers, packed columns, cyclone scrubbers, and venturi scrubbers are employed to absorb pollutant gases. Absorption is used extensively in the separation of corrosive and hazardous pollutants from waste gases. These technologies are explained above.

Spray towers can handle a fairly large volume of gas with relatively low pressure drop and reasonably high efficiency of removal. Spray towers are also effective for dual removal of particulate and gaseous contaminants as explained earlier.

Adsorption

Generally, absorbed materials are dissolved into the absorbent, like sugar in water, whereas adsorbed materials are attached onto the surface of a material, like dust on a wall.

This section deals with adsorption, which is also a mass transfer process in which the gas is bonded in a solid. Adsorption is a surface phenomenon. The mechanism of adsorption can be classified as either physical adsorption or chemisorption. The bond may be physical or chemical. Electrostatic force (van der Walls) holds the pollutant gas when physical bonding is significant" [Raj 08].

NOTE *Van derWalls forces or interactions are the total forces of attraction and repulsion between molecules other than covalent bonds.*

"Chemical bonding is by reaction with the surface. Activated carbon, molecular sieves, silver gel, and activated alumina are the most common adsorbents. The common property of the adsorbent is a large surface area per unit volume. They are very effective for hydrocarbon pollutants. This process is particularly suitable when the pollutants are:

- Noncombustible
- Unstable in liquid
- Low in concentration

Condensation

A compound will condense at a given temperature if the partial pressure is increased until it is equal to or greater than its vapor pressure at that temperature. If the temperature of the gas mixture is reduced to its saturation temperature, its vapor pressure equals its partial pressure and condensation will occur.

We can remove VOCs from the gas stream by cooling them to a lower temperature, so that most of the VOCs are condensed as liquid and then separated from the gas by gravity.

Biofiltration

The biological treatment of VOCs and other pollutants has received increasing attention in recent years. Biofiltration involves the removal and oxidation of organic compounds from contaminated air by beds of compost, peat, or soil. This treatment offers an inexpensive alternative to conventional air treatment technologies such as adsorption and incineration.

The schematic representation of biofiltration is shown in Figure [6.7]. A biofiltration process uses microorganisms immobilized in the form of a biofilm layer on adsorptive materials such as compost, peat, or soil. As the contaminated vapor streams pass through the filter bed, pollutants are transferred from the gaseous phase to the liquid biolayer and oxidized. Biofilters can be used for treating vapor containing about 1,500 MG/MG^3 of biodegradable VOCs" [Raj 08].

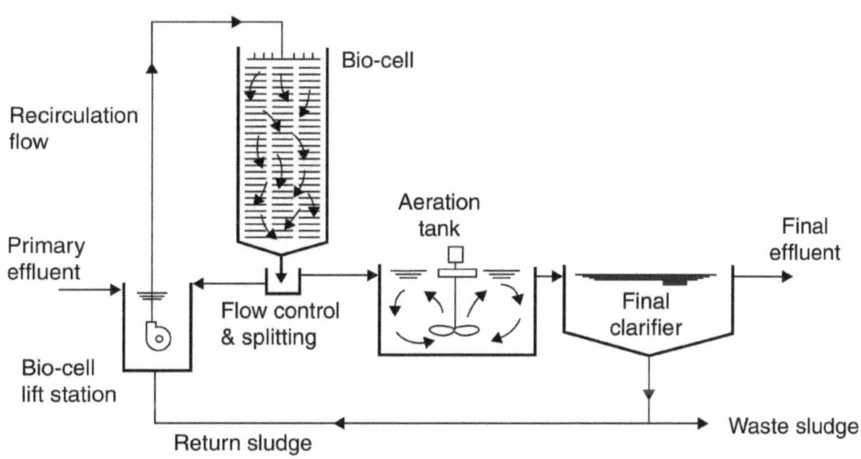

FIGURE 6.7 Schematic representation of biofilter. (Adapted from EPA-600/S2-82-067.)

Combustion

"Some air pollutants contain materials such as compounds of carbon, hydrogen, oxygen, and sulfur which, when burned, produce less harmful chemicals.

Some examples are:

$$CO + 1/2\ O_2 \rightarrow CO_2$$
$$C_6H_6 + 7\ 1/2\ O_2 \rightarrow 6CO_2 + 3H_2O$$
$$H_2S + 3/2\ O_2 \rightarrow H_2O + SO_2$$

Carbon monoxide (CO), which is harmful to health, converts to carbon dioxide. Benzene, which is a reactive hydrocarbon, is converted to carbon dioxide and water.

The combustion equipment used to control pollution emission is designed to promote oxidation reactions as close as possible to completion, leaving a minimum of unburned compounds. For proper combustion to occur, it is necessary to have a proper combination of four basic parameters, such as oxygen, temperature, turbulence, and time. Depending upon the contaminant being oxidized, the combustion method can be classified as:

- Direct flame combustion
- Thermal combustion
- Catalytic combustion

Direct Flame Combustion

In direct flame combustion, the waste gases are burned directly in an incinerator, with or without addition of supplementary fuel. In some cases, heat value and oxygen content are sufficient to allow them to burn on their own. In other cases, introducing air and adding a small amount of supplementary fuel will bring the gaseous mixture to its combustion point.

Thermal Combustion

Thermal incinerators are used when the concentration of combustible gaseous pollutants are too low to make a direct flame combustion feasible. Generally the waste gas is preheated and the preheated gas is directed into a combustion zone, equipped with a burner supplied with supplementary fuel. The temperature of the operation depends upon the nature of the pollutants in the waste gas.

Catalytic Combustion

A catalytic combustion unit generally consists of a preheating section and catalytic section [Figure 6.8]. The catalytic combustion process has been used to control the formation of SOx, NOx, HC, and CO. In the catalytic removal of SO_2, the gas is thoroughly cleaned in a dust precipitator, then passed through an SO_2 oxidation catalyst, vanadium pentoxide, at high temperature (454 °C). The process yields sulfuric acid mist. The process has an estimated capacity for removing 90% of the original SO_2 present in the flue gas" [Raj 08].

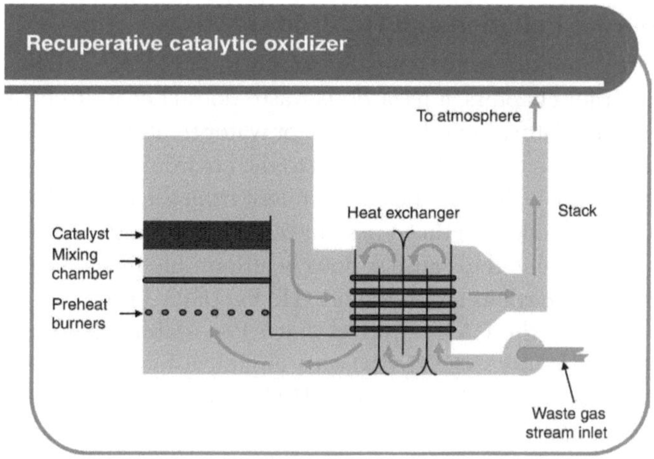

FIGURE 6.8 Schematic representation of a catalytic incinerator. (Adapted from EPA Technology Transfer Network. *http://cfpub.epa.gov/oarweb/mkb/contechnique.cfm?ControlID=10.*)

6.7 EPA PERFORMANCE STANDARDS FOR INCINERATORS

The Subpart O standards for hazardous waste incinerators set performance standards that limit the quantity of gaseous emissions an incinerator may release. Specifically, the regulations set limits on the emission of organics, HCl, and PM. The following section outlines the requirements for each of these substances.

Organics

"To obtain a permit, an owner/operator must demonstrate that emission levels set for various hazardous organic constituents are not exceeded. EPA's principle measure of incinerator performance is its destruction and removal efficiency (DRE). A 99.99 percent DRE means that one molecule of an organic compound is released to the air for every 10,000 molecules entering the incinerator. A 99.9999 percent DRE means that one molecule of an organic compound is released to the air for every one million molecules entering the incinerator.

Since it would be impossible to monitor the DRE results for every organic constituent contained in a waste, certain principal organic hazardous constituents (POHCs) are selected for monitoring and are designated in the permit. POHCs are selected based on high concentration in the waste feed and difficulty in burning compared to other organic compounds. If the incinerator achieves the required DRE for the selected POHCs, then it is presumed that the incinerator should achieve the same or better DRE for organic compounds that are easier to incinerate. RCRA performance standards require a minimum DRE of 99.99 percent for POHCs designated in the permit and a minimum destruction and removal efficiency of 99.9999 percent for dioxin-bearing wastes F020, F021, F022, F023, F026, or F027 (Section 264.343(a))" [EPA 11l].

Relationship Between Air Pollution and Hazardous Waste

Hazardous waste combustors are regulated jointly under RCRA and the Clean Air Act (CAA). As mentioned in earlier chapters, a hazardous waste doesn't cease to be a hazardous waste just because it crosses a regulatory boundary into air or water treatment systems. These materials are not hazardous wastes when they are discharged to the environment. In the case of air discharges, the CAA protects human health and the environment from the harmful effects of air pollution by requiring significant reductions in the emissions of the most dangerous air pollutants. These pollutants are known or suspected to cause serious health problems such as cancer or birth defects, and are referred to as hazardous air pollutants (HAPs) [EPA 11m].

As originally enacted, the CAA required that EPA establish National Emission Standards for Hazardous Air Pollutants (NESHAPs) on a chemical-by-chemical basis. Under this mandate, EPA established NESHAPs for seven HAPs. However, the 1990 amendments to the CAA changed EPA's approach to regulating HAPs, so that NESHAPs are now established based on the 'maximum achievable control technology' (MACT) for an industry group or source category (for example, hazardous waste combustors). These standards are based on emission levels that are already being achieved by the better-performing sources within the group.

The NESHAP for hazardous waste combustors was developed in two phases. Phase I, which addressed hazardous waste burning incinerators, cement kilns, and lightweight aggregate kilns, were originally promulgated on September 30, 1999 (64 FR 52828). Hazardous waste burning industrial boilers, process heaters, and hydrochloric acid production furnaces, were addressed in Phase II, which was signed on September 14, 2005. Replacement standards for Phase I also were signed on this date [EPA 11m].

Because this text is dedicated to hazardous waste management, the details of these requirements are not covered. However, they can be researched online at *www.epa.gov.*

The following case study illustrates a part of the efforts that must be undertaken by owners/operators of hazardous waste incinerators/combustors who wish to be permitted to incinerate/combust hazardous waste.

Case Study

Trial Burns for Hazardous Waste Combustors

As mentioned in the beginning of the chapter on hazardous waste treatment (Chapter 4, "Hazardous Waste Treatment and Disposal"), hazardous wastes do not cease to be hazardous when they are discharged to a device or system that is regulated under another regulatory program (air or water). They do cease to be a hazardous waste when they are released to the environment after treatment.

Hazardous waste incinerators, boilers, and industrial furnaces are regulated and permitted under the hazardous waste and air resources regulatory program. In order for a hazardous waste incinerator to renew or maintain a hazardous waste permit and an air pollution control permit to combust or incinerate hazardous wastes, the incinerator must periodically prove it can achieve the Destruction Removal Efficiency (DRE) and other air discharge requirements in the USEPA regulations.

After the facility prepares a waste analysis plan and numerous design documents to propose waste feed rates and air mixture ratios, the facilities must undergo test or trial burns to prove the incinerator can operate effectively under less than ideal conditions. These trial burns can last several days, and are very expensive. Regulatory staff observe incinerator operators and testing company staff while premeasured and mixed waste streams are fed to the incinerator, under varying conditions. The waste feed rates, operating conditions, and stack emissions are all monitored carefully during these trial burns, and the results determine if the incinerator can handle the proposed waste feed rates without subjecting the public or the environment to undue risks.

6.8 AUTOMOBILE POLLUTION

The following segment is written as a practical illustration of hazardous waste combustion, with the fuel (gasoline) representing the waste, and the vehicles as pollution control facilities. Actual air pollution control facilities are much more elaborate than vehicles, as they are required to capture the more dangerous and substantial products of hazardous waste combustion.

Automobiles are a major source of man-made air pollution in the United States today. A discussion of these emissions is as follows:

Classification of Emissions

"The emissions produced by a vehicle fall into two basic categories:

Tailpipe Emissions

The products of burning fuel in the vehicle's engine emitted from the vehicles exhaust system. The major pollutants include:

- **Hydrocarbons**: unburned or partially burned fuel;
- **Nitrogen oxides (NOx)**: These are generated when nitrogen in the air reacts with oxygen under the high temperature and pressure conditions inside the vehicle's engine. NO_x emissions contribute to both smog and acid rain.
- **Carbon monoxide (CO)**: A product of incomplete combustion, carbon monoxide reduces the blood's ability to carry oxygen by replacing the oxygen in the human body, and is very dangerous to people with heart disease.

Evaporative Emissions

These are produced from the evaporation of fuel. Fuel tends to evaporate in these ways:

- **Gas tank venting**: The heating of the vehicle's fuel tank as the temperature rises from the night-time temperature to the hottest temperature of the day means that gasoline in the tank evaporates, increasing the pressure inside the tank above atmospheric pressure. To relieve this pressure, the gasoline fumes are vented to the atmosphere.
- **Running losses**: The escape of gasoline vapors from the hot engine.

- **Refueling losses**: The empty space inside a vehicle's fuel tank is filled with hydrocarbon gases, and as the tank is filled, these gases are displaced and forced out of the filler pipe into the atmosphere. In addition, there is loss from further evaporation and fuel spillage" [Raj 08].

Factors Influencing Emissions

"The emissions from a vehicle depend on a wide range of factors:

Fuel Used

Emissions depend on the fuel used to power the vehicle. For example, a car powered by gasoline will emit more carbon monoxide (CO) and VOCs, and be less fuel efficient than a similar one powered by diesel. However, the diesel vehicle will emit more NOx and particulates than the gasoline vehicle. Carbon dioxide (CO_2) emissions depend on fuel consumption and the carbon content of the fuel.

Maintenance

Poorly maintained vehicles consume more fuel and emit higher levels of carbon monoxide (CO) and VOCs than regularly serviced ones.

Driver Behavior

Emissions are not constant, but vary depending on how the vehicle is being driven. NOx emissions increase when the engine is under load such as during rapid acceleration and when travelling at high speeds. Carbon monoxide (CO) and VOC emissions will increase when the vehicle is running rich, for example, when the engine is cold and during accelerations. Thus, in general, emissions will be lowest when a car is accelerated slowly and driven at a steady speed. In the stop-start driving conditions that characterize congested urban areas, emissions will be higher than at the same average speed but under free flow (steady speed) conditions.

Cold Starts

Most car journeys are relatively short and are in urban areas. Emissions from cars are particularly high when first driven after a cold start" [Raj 08].

Control Measures

Optimizing Air-Fuel Ratio

"In a typical gasoline-fueled automobile engine, with no air pollution controls, a mixture of gasoline and air is fed into a cylinder by a carburetor or fuel injector and is compressed and ignited by a spark from the spark plug. The explosive energy of the burning mixture moves the pistons. The force of the piston's motion is transmitted to the crankshaft that drives the car.

One kilogram of gasoline can burn completely when mixed with about 15kg of air. For maximum power, proportion of air to fuel (A/F) must be less. The A/F ratio is fairly easy to regulate, and has been refined greatly by the use of computers. The A/F ratio has a direct effect on the emission of CO, HC and NOx. A/F ratio of 14.6 is the stoichiometric mixture for complete combustion. At low ratios, both CO and HC emissions increase. At very lean mixtures (high A/F ratios), the NOx emission begins to decrease. Thus, one of the approaches taken to control emission is to set the carburetor or fuel injector at a very lean setting (see Figure [6.9]).

NOTE *As stated earlier, the entire discussion under Section 6.8 Automobile Pollution is for illustrative purposes, and is in direct correlation with the current principles of hazardous waste incineration. The fuel issues are similar, the emissions are similar (depending on the fuel), as are the pollution controls, such as air-fuel ratio, and control of evaporative emissions. One of the key differences is that incinerator pollution controls often have to deal with other pollutants in the waste fuel, like dust, metals, and other products of incomplete combustion.*

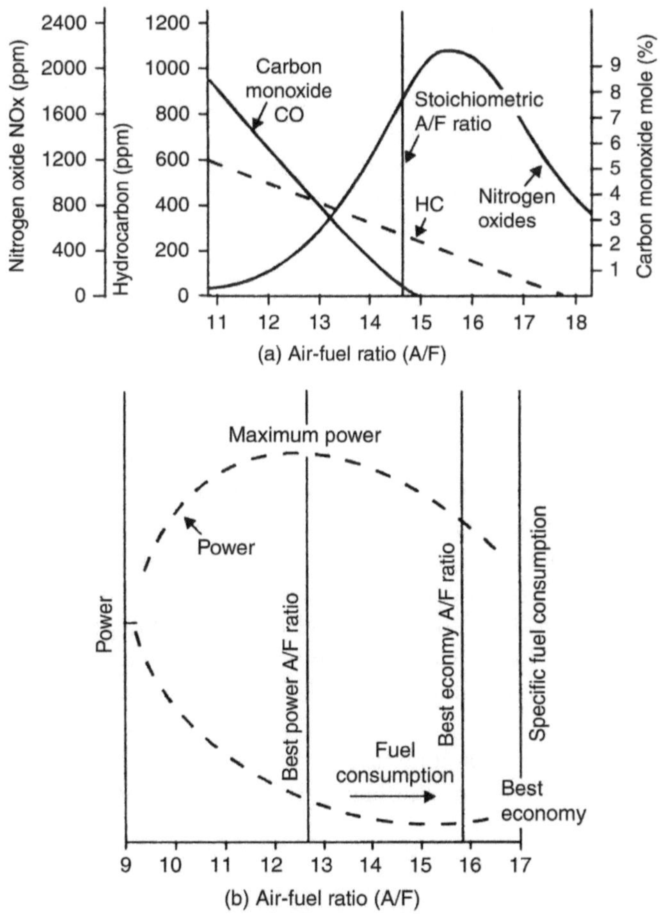

FIGURE 6.9 Effect of air-fuel ratio (a) on emissions (b) on power and economy. (Permission granted courtesy of Laxmi Publications Pvt. Ltd.)

Catalytic Converter

The greatest step so far in controlling emissions from road vehicles was the introduction of closed loop (or controlled) three-way catalysts for gasoline vehicles. These remove 80–90% of the emissions of CO, VOCs, and NOx. The reactions taking place on the catalyst are shown below:

- Oxidation reactions:
  - $2CO + O_2 \rightarrow 2CO_2$
  - $HC + O_2 \rightarrow CO_2 + H_2O$
- Reduction reactions:
  - $2CO + 2NO \rightarrow 2CO_2 + N_2$
  - $HC + NO \rightarrow CO_2 + H_2O + N_2$

Automotive catalysts are typically made of platinum and rhodium. For efficient removal of all three pollutants, the A/F ratio needs to be close to the stoichiometric ratio (i.e., 14:7). Cars fitted with these catalysts require an oxygen sensor to monitor the exhaust gas composition and electronically controlled fuel management system to control the A/F ratio.

This technology cannot be used in the oxygen-rich exhaust of a lean burn gasoline or diesel engine. In these engines, CO and HC emissions can be reduced using a simple oxidation catalyst" [Raj 08].

NOTE *The Stoichiometric Ratio can be defined as the optimal mix of fuel and air to achieve complete combustion of the mixture.*

Control of Evaporative Emissions

"The HC-containing vapors from the carburetor or fuel injector, and the fuel tank pass through an activated carbon bed that removes the HC before those vapors are vented to the air. When the engine is running at other than idle, or very low speeds, air is sucked through the bed in the reverse direction, removing the HC from the activated carbon, preparing it for its next service. This regenerated air is returned to the air intake of the engine, where the HC is mixed with the fresh air coming in and is burned. Suitable valves maintain the flow in the proper direction. The activated carbon bed has a low enough flow resistance that the pressure in the carburetor or fuel injector is close enough to atmospheric pressure for proper operation.

Needless to say, the advent and use of fuel injection revolutionized the automobile industry, making the fuel mixture automatically set and adjusted by computerized controls" [Raj 08].

6.9 CASE STUDIES

The case study below presents a dilemma for industries and regulators alike. The complaints by the neighbors concerning the new and safer solvent forced the company to revert to the use of the original, more toxic and ozone-depleting solvent.

Case Study

Complaints by Neighbors Cause More Hazardous Waste and Air Pollution

A hazardous waste inspector was called out on an anonymous odor complaint from a member of the public. Upon investigation, there were nuisance odors emanating from a small company that produced precision ceramics used for casting of metal machinery parts. Upon arriving at the site, the inspector smelled a strong citrus odor, similar to oranges.

The inspector inquired about the smell, and the owner/operator explained as follows: The company had been using a very low molecular weight (1, 1, 1 trichloroethylene) as a solvent to release the ceramic molds from their casts. When the owner researched the issue of using this solvent, he discovered that 1, 1, 1 trichloroethylene, as a halogenated compound, was shown to deplete the Earth's ozone layer; he switched to a nonhazardous, citrus-based solvent. By the time the inspector had arrived, the facility had already tried several citrus fruit-based solvents, including lemons, limes, and grapefruit. While all of these solvents were effective for their intended use (releasing the molds), the neighbors complained about the odors. The facility was trying an orange-based solvent as a last attempt, and that was generating complaints too.

It was clear that the 1, 1, 1 trichloroethylene was evaporating into the air more rapidly than the citrus-based solvents, and the neighbors never smelled it, so they didn't complain. Since the citrus-based solvents were higher molecular weight, their odors did not evaporate as quickly; the residents could smell the odors and complained.

This facility continued their search for a nonhazardous, non-ozone depleting solvent that wouldn't generate complaints, but over the short term returned to using the low molecular weight solvent.

This is an example of how waste reduction isn't always as effective as planned.

This next case study illustrates that compliance with federal, or even state, laws is not enough, because the local laws, rules, and regulations may be even more stringent.

Case Study

Dry Cleaners Driven by Regulations

In Chapter 1, "A Brief History of Hazardous Waste," there is a discussion of the hazardous waste Land Disposal Restriction (LDR) program, and how these regulations prescribe treatment standards for every waste, sometimes even prescribing specific treatment technologies. A similar approach has been used in the federal air resources program, where, starting in September 1993, dry cleaners using perchloreylene (perc) are required to install certain "generations" of design by prescribed deadlines, particularly for residential buildings.

EPA's goal is to phase out the use of perc in residential buildings by 2020. They encourage the use of alternative technologies (CO_2, hydrocarbon, and wet cleaning). A useful and comprehensive multimedia compliance checklist was developed by EPA that helps dry cleaning establishments understand the regulatory requirements of all EPA programs. It is highly advisable for all industries to look at their facility's processes and emissions from a broad perspective, assuring optimum environmental protection across all programs, often discovering ways to save money and optimize production in various processes.

The final case study in this chapter is about a facility that routinely burned tons of pressure-treated wood chips, creating poisonous gases for years before a hazardous waste inspector discovered the violation.

Case Study

Facility Caught Burning Pressure-Treated Wood

A hazardous waste inspector was assigned to inspect a facility that made structures out of wood products. When the inspector arrived, the facility operator was very cooperative, and showed the inspector the entire facility. During the walk-through, the inspector noticed that the facility collected all of their sawdust and wood chips in a vacuum system. The operator explained that they collected all the wood chips and sawdust and burned it in a wood boiler to heat the facility.

When the inspector noticed the facility had some copper/chrome/arsenic (CCA) pressure-treated wood in the building, he asked if the facility also burned that wood in the boiler. The facility operator responded "yes," because it was difficult to separate the plain wood from the CCA wood. The inspector asked if they had ever done a hazardous waste determination on the CCA wood, and the operator responded no.

The inspector explained that burning CCA wood was prohibited, because the combustion releases the metals into the air, including poisonous arsine gas. The facility immediately stopped burning the CCA wood, and came up with a way to separate the CCA wood from the untreated wood. Testing of the CCA wood chips proved to be hazardous for chromium approximately 50% of the time.

Fortunately, this facility was in a rural area, so no known human poisoning was reported, but the facility had to prove the separation process was effective, and paid for disposal of the CCA wood at a licensed hazardous waste facility instead of benefiting from the heat value by burning it. In the end the facility remained economically viable, but the prospect of releasing the poison arsine gas was stopped.

The information provided in this chapter is a small portion of the work facilities must do to treat their air discharges, including the hazardous air pollutants. The tests are rigorous and are performed at great expense, with no guarantee of success. It is not uncommon for incinerators to conduct several trial burns before they finally prove their facility can be operated with no significant risk to human health or the environment.

Although we briefly mentioned the challenges faced by dry cleaners in a case study, the regulatory requirements for auto-body shops, dry cleaners, and many other small industries are very expensive, and cut into company profits. Many of these companies are finding that source reduction, waste minimization, and product substitution cut their waste generation and disposal costs significantly.

The cost savings drive these facilities to carefully examine their manufacturing and waste management practices, and are changing the way many of these firms do business.

In this chapter, you learned the history of air pollution in the United States, the structure of the atmosphere, the sources of air pollutants, the classification of air pollutants, the effects of air pollution, and air pollution control technologies. You read about performance standards for incinerators, studied automobile pollution and control technologies, and learned the relationship between air pollution control and hazardous waste. You also read case studies on complaints by neighbors causing more hazardous waste and air pollution, dry cleaners driven by regulations, and trial burns for hazardous waste combustors.

In the next chapter, you will learn about U.S. wastewater regulations, learn the sources of contamination of wastewaters, how to classify water pollutants, how to characterize wastewater, discover different kinds of wastewater treatment, read a summary of the U.S. storm water regulations, read a case study on illegal wastewater treatment. Finally, you will read an article from the field on how to start treating a difficult wastewater.

EXERCISES

1. From a regulatory perspective, does a hazardous waste cease to be hazardous when it is discharged to an air pollution control system that is regulated under the Clean Air Act? Why?
2. What is air pollution?
3. Name the two main sources of air pollution.
4. What is the smallest of all the particle sizes of pollutants listed in physical characteristics?
5. What are two broad effects of air pollution?
6. Name three devices used to control particulate matter.
7. Name five ways to control gaseous contaminants.
8. Name three ways gasoline-powered vehicles cause air pollution.
9. What has been the greatest step so far in controlling tailpipe emissions?
10. Explain how a vehicle's carbon canister controls fuel and vapor losses.

REFERENCES

[EPA11k] USEPA Air Pollution Control Orientation Course Origins of Modern Air Pollution Control Regulations, online at *http://www.epa.gov/apti/course422/apc1.html*, (accessed May 20, 2011).

[EPA 11l] EPA RCRA Hotline Training 530-R-99-052 PB2000-101 892 online at *http://www.epa.gov/osw/inforesources/pubs/hotline/training/incin.txt*, (accessed May 20, 2011).

[EPA 11m] USEPA NESHAPS Tool Kit, online at *http://www.epa.gov/osw/hazard/tsd/td/combust/toolkit/index.htm*, (accessed May 20, 2011).

[Raj 08] Raj, S. Amal. 2008. Introduction to environmental science and technology. 27–41. Laxmi Publications Pvt, Ltd.

CHAPTER 7

Wastewater Management

In This Chapter

- U.S. wastewater regulations
- Sources of contamination of wastewaters
- Classification of water pollutants
- Characteristics of wastewater
- Wastewater treatments
- U.S. storm water regulations
- Case study and field article

Hazardous waste regulations apply to the storage, transportation, and treatment of hazardous waste, but in general terms; these regulations do not apply when the waste is discharged to an air pollution control system or a wastewater treatment system. That does not mean the waste ceases to be hazardous when discharged to another system; it means the discharge has crossed a "regulatory boundary." Air discharges are regulated under the Clean Air Act and waste water discharges are regulated under the Clean Water Act.

This chapter provides basic information on the treatment of wastewater as it pertains to hazardous waste. These sources are primarily from industries and wastewater treatment plants.

First, let's discuss the regulations governing wastewater.

Numerous portions of this chapter have been quoted directly with permission from Dr. S. Amal Raj, *Introduction to Environmental Science and Technology*. Laxmi Publications Pvt. Ltd. 2008. 42-56.

7.1 U.S. WASTEWATER REGULATIONS

The first major law governing water pollution was the Federal Water Pollution Control Amendment of 1948, passed as a measure to govern and control water pollution. Significant amendments were introduced in the Clean Water Act, passed on October 18, 1972, which set up the National Pollutant Discharge Elimination System (NPDES), creating permits for point sources of pollution. These permits applied to municipal and industrial wastewater discharges [WIKI 11f].

"The 1972 Amendments to the Federal Water Pollution Control Act (Public Law 92-500), known as the Clean Water Act (CWA), established the foundation for wastewater discharge control in this country. The CWA's primary objective is to 'restore and maintain the chemical, physical and biological integrity of the nation's waters.'

The CWA established a control program for ensuring that communities have clean water by regulating the release of contaminants into our country's waterways. Permits that limit the amount of pollutants discharged are required of all municipal and industrial wastewater dischargers under the National Pollutant Discharge Elimination System (NPDES) permit program. The NPDES permit process relies on self-monitoring and reporting by the permit holder vis Discharge Monitoring Reports (DMRs). In addition, a construction grants program was set up to assist publicly owned wastewater treatment works build the improvements required to meet these new limits.

The 1987 Amendments to the CWA established State Revolving Funds (SRF) to replace grants as the current principal federal funding source for the construction of wastewater treatment and collection systems.

Over 75 percent of the nation's population is served by centralized wastewater collection and treatment systems. The remaining population uses septic or other onsite systems. Approximately 16,000 municipal wastewater treatment facilities are in operation nationwide. The CWA requires that municipal wastewater treatment plant discharges meet a minimum of 'secondary treatment.' Over 30 percent of the wastewater treatment facilities today produce cleaner discharges by providing even greater levels of treatment than that offered by secondary treatment" [EPA 04, p.4].

NOTE *"Water pollution can be defined as any alteration in physical, chemical, or biological properties of water, rendering the water harmful to public health and the environment" [Raj 08].*

Nonpoint sources, such as storm water runoff and runoff from agricultural areas, were not addressed in the 1972 amendments. Although this text does not deal with nonpoint sources in detail, an overview of the various categories of nonpoint sources is given in Section 7.3. A brief but useful and informative history of the federal storm water regulatory program is summarized in Section 7.6 of this chapter.

7.2 SOURCES OF CONTAMINATION OF WASTEWATER

"For convenience, the sources of contamination of water can be classified as natural and anthropogenic (man-made). [We] will not deal with natural sources, since the subject is hazardous waste, which is entirely man-made" [Raj 08].

Anthropogenic (Man-Made) Sources

"Man-made sources of wastewater are the result of industrial, domestic, agricultural, and mining activities of humans, and can be characterized as follows:

- **Industrial sources:** The discharges from most industries are subject to industrial pretreatment requirements where necessary, and then discharged to a publicly owned treatment works (POTW) or discharged directly to surface water under a NPDES permit. Examples of industries that generate fairly large quantities of wastewater are: pulp and paper mills, tanneries, and chemical manufacturing facilities.
- **Domestic sources:** For the most part, domestic wastewater comes from residences, commercial offices buildings, schools, and other institutions. There can be hazardous constituents in these wastewaters, such as household cleaners, laboratory wastes from schools, and household hazardous wastes dumped down the sewer.
- **Agricultural sources:** can include soil and silt from erosion, agricultural run-off, and synthetic fertilizers, herbicides, and insecticides. Out of these pollutants, only the pesticides (herbicides and insecticides) might be hazardous wastes. One of the problems with several synthetic pesticides, particularly the chlorinated hydrocarbons (DDT, endrin, chlordane, etc.) is that they are persistent in the environment, meaning they are resistant to degradation, so they stay in the environment in their original form for a very long time.
- **Mining activities:** Mining operations can produce soluble toxic material, acid drainage, and silt and other *fines* in runoff" [Raj 08].

NOTE *Fines refers to the finely crushed or powdered particles of ore found in a mixture of particulates of varying sizes.*

7.3 CLASSIFICATION OF WATER POLLUTANTS

"To fully understand the effects of water pollution and the technologies applied to its control, it is useful to classify pollutants into various categories. First, as mentioned earlier, a pollutant can be classified according to its nature of origin, as either a *point* source or a *nonpoint* (dispersed) source" [Raj 08].

Point Sources

The treatments of discrete discharges from pipes and other conduits from identifiable point sources, such as manufacturing plants, businesses, industries, and wastewater treatment plants serving all of these industries, together with residences, are discussed in other chapters of this text. To briefly review, examples of nonresidential point sources include: chemical manufacturing facilities, tanneries, electroplating facilities, auto body shops, laboratories, and mining operations. In this section, we will deal primarily with nonpoint sources.

Nonpoint Sources

Some of the nonpoint sources of concern in the hazardous waste field are:

Oxygen-Demanding Substances

"Dissolved oxygen is a key element in water quality that is necessary to support aquatic life. A demand is placed on the natural supply of dissolved oxygen by many pollutants in wastewater.

This is called biochemical oxygen demand, or BOD, and is used to measure how well a sewage treatment plant is working. If the effluent, the treated wastewater produced by a treatment plant, has a high content of organic pollutants or ammonia, it will demand more oxygen from the water and leave the water with less oxygen to support fish and other aquatic life.

Organic matter and ammonia are 'oxygen-demanding' substances. Oxygen-demanding substances are contributed by domestic sewage and agricultural and industrial wastes of both plant and animal origin, such as those from food processing, paper mills, tanning, and other manufacturing processes. These substances are usually destroyed or converted to other compounds by bacteria if there is sufficient oxygen present in the water, but the dissolved oxygen needed to sustain fish life is used up in this break down process" [EPA 04, p.8].

Pathogens

Pathogens are agents that cause disease, like bacteria, fungi, and viruses. "Disinfection of wastewater and chlorination of drinking water supplies has reduced the occurrence of waterborne diseases such as typhoid fever, cholera, and dysentery, which remain problems in underdeveloped countries while they have been virtually eliminated in the U.S.

Infectious micro-organisms, or pathogens, may be carried into surface and groundwater by sewage from cities and institutions, by certain kinds of industrial wastes, such as tanning and meat packing plants, and by the contamination of storm runoff with animal wastes from pets, livestock and wild animals, such as geese or deer. Humans may come in contact with these pathogens either by drinking contaminated water or through swimming, fishing, or other contact activities. Modern disinfection techniques have greatly reduced the danger of waterborne disease.

Nutrients

Carbon, nitrogen, and phosphorus are essential to living organisms and are the chief nutrients present in natural water. Large amounts of these nutrients are also present in sewage, certain industrial wastes, and drainage from fertilized land. Conventional secondary biological treatment processes do not remove the phosphorus and nitrogen to any substantial extent—in fact, they may convert the organic forms of these substances into mineral form, making them more usable by plant life. When an excess of these nutrients over stimulates the growth of water plants, the result causes unsightly conditions, interferes with drinking water treatment processes, and causes unpleasant and disagreeable tastes and odors in drinking water. The release of large amounts of nutrients, primarily phosphorus but occasionally nitrogen, causes nutrient enrichment which results in excessive growth of algae. Uncontrolled algae growth blocks out sunlight and chokes aquatic plants and animals by depleting dissolved oxygen in the water at night [EPA 04, p.8].

NOTE *"The release of nutrients in quantities that exceed the affected water body's ability to assimilate them results in a condition called eutrophication or cultural enrichment" [EPA 04, p.8].*

Inorganic and Synthetic Organic Chemicals

A vast array of chemicals is included in this category. "Examples include detergents, household cleaning aids, heavy metals, pharmaceuticals, synthetic organic pesticides and herbicides, industrial chemicals, and the wastes from their manufacture. Many of these substances are toxic to fish and aquatic life and many are harmful to humans. Some are known to be highly poisonous at very low concentrations. Others can cause taste and odor problems, and many are not effectively removed by conventional wastewater treatment.

Thermal

Heat reduces the capacity of water to retain oxygen. In some areas, water used for cooling is discharged to streams at elevated temperatures from power plants and industries. Even discharges from wastewater treatment plants and storm water retention ponds affected by summer heat can be released at temperatures above that of the receiving water, and elevate the stream temperature. Unchecked discharges of waste heat can seriously alter the ecology of a lake, a stream, or estuary" [EPA 04, p.8].

Fertilizers and Other Chemicals

"Chemicals that provide nutrients to plants, such as nitrogen and phosphorous and compounds containing these nutrients, may result from agricultural runoff and sewage treatment plant effluents. These nutrients stimulate the growth of aquatic plants, which interfere with water uses and later decay to add biological oxygen demand (BOD) to water.

Sediments

The natural process of soil erosion gives rise to sediments in water. Sediments are particles and other matter from eroded soil, sand, and minerals. Rivers have always carried sediments to the oceans, but erosion rates in many areas have been greatly accelerated by human activities. In general, sediments contain soil and mineral particles that are washed from the land, agricultural fields, forest, grazing lands, and construction sites. As described briefly above, the U.S. Congress set up a federal storm water program (MS4) to require municipalities to address these problems" [Raj 08].

Radioactive Pollutants

The discharge of uranium, thorium, and other radioactive materials from hospitals and laboratories contaminate the surface water and remain a problem until they are removed or until they naturally decay, which can sometimes take a very long time.

7.4 WASTEWATER CHARACTERIZATION

"The characterization of wastewater is necessary to find out the various types of contaminants present in the wastewater, along with their concentration. This information will help in identifying the specific pollutants and the type of treatment necessary before disposal.

Characteristics of waste water are:

- **Physical**, as in color, odor, turbidity, temperature, and solids content
- **Chemical**, as in pH, alkalinity, inorganic constituents like chlorides, heavy metals, nitrogen, phosphorous, etc., dissolved oxygen, biochemical oxygen demand (BOD), and chemical oxygen demand (COD)
- **Biological**, as in bacteria, algae, protozoa, viruses, and coliforms.

Some physical, chemical, and biological parameters are described below:

- **pH:** An expression of both acidity and alkalinity on a scale of 0–14, with 7 representing neutrality. Numbers less than 7 indicate increasing acidity, and numbers higher than 7 indicate increasing alkalinity. pH is expressed on a logarithmic scale, so each whole pH value above and below 7 is ten times more basic or acidic than the next higher value.
- **Alkalinity:** Alkalinity in waste water results from the presence of hydroxides, carbonates, and bicarbonates of elements such as calcium, magnesium, sodium, potassium, or ammonia. The alkalinity in wastewater helps to resist changes in pH caused by the addition of acids.
- **Dissolved oxygen:** The amount of oxygen freely available in water and necessary for aquatic life and the oxidation of organic materials.
- **Oxygen demand:** Chemical and biological oxygen demand (COD and BOD) are measures of the oxygen consumed when a substance degrades. Materials such as food waste and dead plant or animal tissue use up dissolved oxygen in water when decomposed through chemical or biological processes.
- **Biological oxygen demand (BOD):** The amount of oxygen required by aerobic biochemical processes to break down the organic matter in water. BOD is a measure of the pollutional strength of biodegradable waste on dissolved oxygen in water.
- **Chemical oxygen demand (COD):** The amount of oxygen utilized in chemical reactions that occur in water as a result of the addition of wastes. COD is a measure of the pollutional strength of organic waste on dissolved oxygen in water.
- **Coliform bacteria:** A group of bacteria used as an indicator of sanitary quality in water. Exposure to these organisms in drinking water causes diseases such as E-coli" [Raj 08].

7.5 WASTEWATER TREATMENT

"The principle objective of wastewater treatment is generally to allow sewage and industrial effluents to be disposed without any danger to human health or damage to the environment. Conventional wastewater treatment consists of a combination of physical, chemical, and biological processes and operations to remove solids, organic matter, chemicals, and sometimes nutrients from wastewater. Wastewater treatment methods can be broadly classified as:

- **Physical unit operation:** The removal of pollutants by physical forces.
- **Chemical unit operations:** The removal of pollutants by addition of chemicals or by chemical reactions.
- **Biological unit operation:** The removal of pollutants by biological activities.

These treatment methods occur in a variety of combinations in wastewater treatment systems, to provide various levels of wastewater treatment" [Raj 08].

The typical flow diagram for wastewater treatment is shown in Figure 7.1.

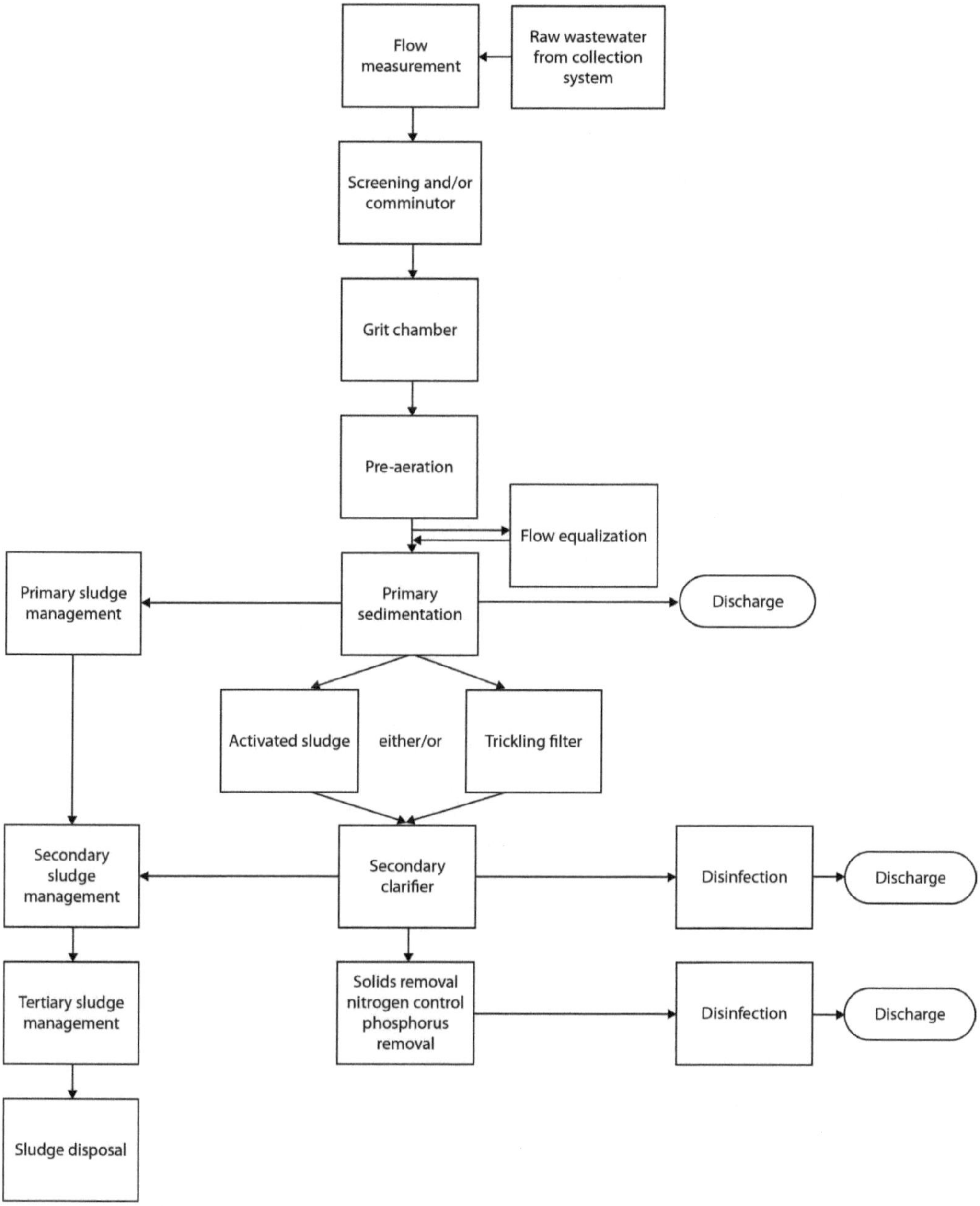

FIGURE 7.1 Possible flow diagram for sanitary wastewater treatment (Note there really isn't a typical industrial treatment process due to the high variability of the contents of industrial wastewater).

Permission granted courtesy of Laxmi Publications Pvt. Ltd.

There are generally four levels of wastewater treatment:

- Preliminary treatment
- Primary treatment
- Secondary treatment
- Tertiary/advanced treatment

Preliminary Treatment

"Preliminary treatment is the first step in wastewater treatment. The purpose of preliminary treatment is the removal of coarse solids and other materials often found in wastewater. This treatment consists mainly of physical unit operations such as:

- **Screening:** The removal of coarse solids in wastewater, which may obstruct or clog the mechanical equipment and pipes. Bar racks are common types of screening devices. Most screens in wastewater treatment plants consist of parallel bars placed at an angle in a channel in such a manner that the wastewater flows through the bars. Trash collects on the bars and is periodically raked off by hand or by mechanical means. In most places, these screenings are disposed of by landfilling or incineration.
- **Comminution (grinding):** The grinding of course solids into smaller and more uniform particles, which are then returned to the flow stream for subsequent treatment.
- **Flotation:** The separation of suspended and floatable solid particles from wastewater. This can be achieved by introducing fine air bubbles into the wastewater.
- **Grit removal:** Grit includes sand, ash, cinder, egg shells, etc., of diameter less than 0.2 mm. The specific gravity of grit varies from 2.0 to 2.6. Grit should be removed early in the treatment process because it is abrasive and rapidly wears out pumps and other equipment. Since it is mostly inorganic, it cannot be broken down by biological treatment processes and thus should be removed as soon as possible" [Raj 08].

Grit is usually removed in a long narrow trench called a "grit channel" (See Figure 7.2)

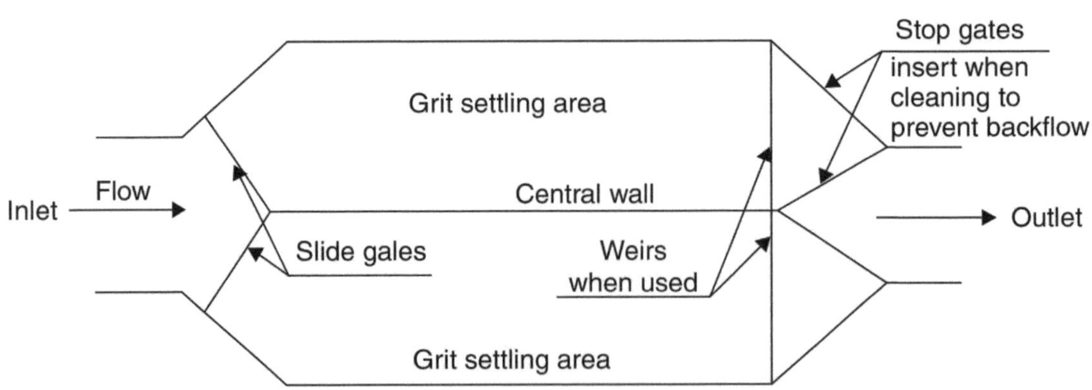

FIGURE 7.2 Grit channel.

(Permission granted courtesy of Laxmi Publications Pvt. Ltd.)

Primary Treatment

"After preliminary treatment, wastewater undergoes primary treatment. The objective of primary treatment is the removal of settleable organic solids by sedimentation and the removal of materials that float (scum) by skimming.

Primary sedimentation tanks (see Figure [7.3]) or clarifiers may be circular or rectangular basins, typically 3 to 5 m deep, with hydraulic retention time (time taken by a particle to travel from inlet to outlet) ranging between two and three hours. In a circular basin, the flow pattern is radial. To achieve the radial flow pattern, the wastewater enters a circular well, designed to distribute the flow equally in all directions. The scraper pushes the settleable solids (sludge) toward the center and into the sludge hopper. The settled solids are known as primary sludge. They are collected for further treatment prior to disposal" [Raj 08].

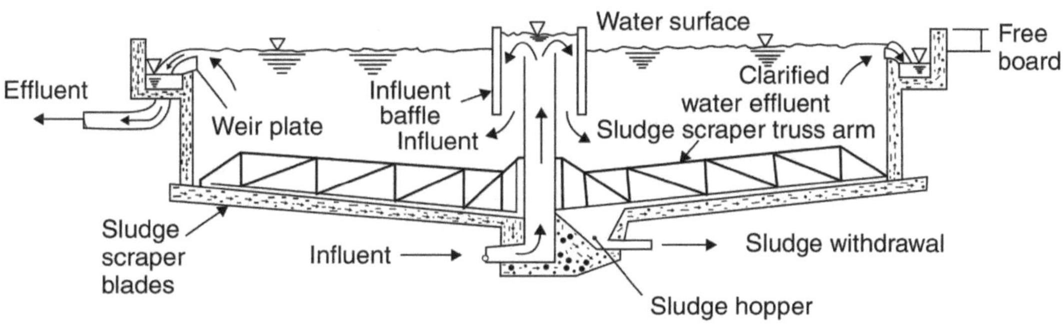

FIGURE 7.3 Sectional view of a circular sedimentation tank.

(Permission granted courtesy of Laxmi Publications Pvt. Ltd.)

"Scum is collected by a rotating blade at the surface. The clear surface water of the primary tank flows from the tank by passing over a weir. The weir must be long enough to allow the treated water to leave at a low velocity. If it leaves the tank at a high velocity, particles settling at the bottom may be picked up and carried from the tank.

Approximately 30% of the incoming biochemical oxygen demand (BOD), 50-70% of the total suspended solids (SS) and 65% of the oil and grease are removed during primary sedimentation. The effluent from primary sedimentation units is called the primary effluent" [Raj 08].

Secondary Treatment (Biological Treatment)

"The goal of all biological treatment systems is to remove the dissolved and nonsettling organic solids from the primary effluent by using microbial populations. Biological treatments are generally part of secondary treatment systems. The microorganisms used are responsible for the degradation of organic matter and the stabilization of organic wastes. The way oxygen is utilized is classified into:

- **Aerobic** (require oxygen for their metabolism)
- **Anaerobic** (grow in the absence of oxygen)
- **Facultative** (can proliferate either in the presence or absence of oxygen).

Stabilization of organic matter by microorganisms in a natural or controlled environment or biological treatment process is accomplished by two distinct metabolic processes:

- **Respiration and Synthesis**
  Respiration is a microbial process in which a portion of the available organic substrate is oxidized by microorganisms to liberate energy. The energy derived from respiration is used to *synthesize* new microbial cells.
  The biological treatment processes used for wastewater treatment are broadly classified as aerobic (in the presence of oxygen) and anaerobic (in the absence of oxygen).

Aerobic Process

Aerobic degradation occurs in two steps. In the first step, complex organics (carbohydrates, proteins, lipids, etc.) are broken down by extracellular enzymes into simple organic compounds. In the second step, aerobic microorganisms (in the presence of oxygen) convert simple organic compounds into oxidized end products such as carbon dioxide, nitrates, and phosphates. The energy released in this process is used for biosynthesis of more bacterial cells" [Raj 08].

Figure 7.4 diagrams the aerobic process.

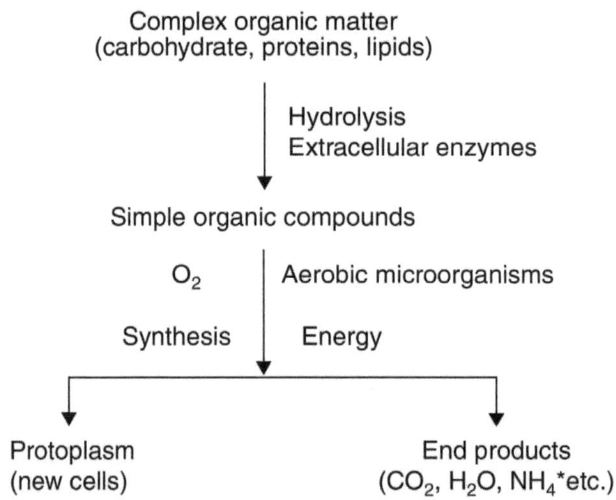

FIGURE 7.4 Microbiology of aerobic process.

(Permission granted courtesy of Laxmi Publications Pvt. Ltd.)

"If the microorganisms are suspended in wastewater during treatment, the operation is called 'suspended growth process.' So the conversion of organic matter to gaseous end products and cell tissues (biomass) can be accomplished aerobically, anaerobically, or facultatively, using suspended and attached growth systems.

Aerobic Biological Treatment Systems

The main aerobic biological wastewater treatment processes include high-rate processes and low-rate processes:

High rate processes are characterized by relatively small reactor volumes and high concentration of microorganisms when compared with low-rate processes. Consequently, the growth rate of new organisms is much greater in high-rate systems, because of the well-controlled environment, and include: activated sludge, oxidation ditch, trickling filter, biofilter (biotower), and rotating biological contactor.

- **Activated sludge process**: a widely-used biological treatment process for both domestic and industrial wastewaters. The activated sludge process refers to a continuous aerobic method for biological wastewater treatment, including carbonaceous oxidation and partial nitrification. The expression 'activated sludge' alludes to a slurry of microorganisms that remove organic compounds from wastewater. These microorganisms are themselves removed by sedimentation under aerobic conditions. In an activated sludge system, soluble and unsettleable, biodegradable organic compounds are degraded by bacteria in an aerated basin, and biomass is carried over with the effluent into a secondary settling tank, where solids are allowed to settle and concentrate; then they are removed. Part of the activated sludge (settled solids) is drawn off as waste, and the rest (30–40%) is recycled to the aeration basin to maintain a constant population of microorganisms" [Raj 08].

NOTE *"This [activated sludge] process originated in England in the early 1900s and earned its name because a sludge (mass of microbes) is produced which aerobically degrades and stabilizes the organic matter of a wastewater" [Raj 08].*

"The process relies on a dense microbial population being mixed in suspension with the wastewater under aerobic conditions. In the presence of adequate nutrients and oxygen, a high rate of microbial growth and respiration is achieved. This results in the utilization of organic matter present, to produce end products such as carbon dioxide, ammonia, phosphate, and sulfate, and biosynthesis of more microorganisms. In activated sludge systems, organic load removals of 85-95% can be achieved" [Raj 08]. Figure 7.5 displays the layout of an activated sludge process.

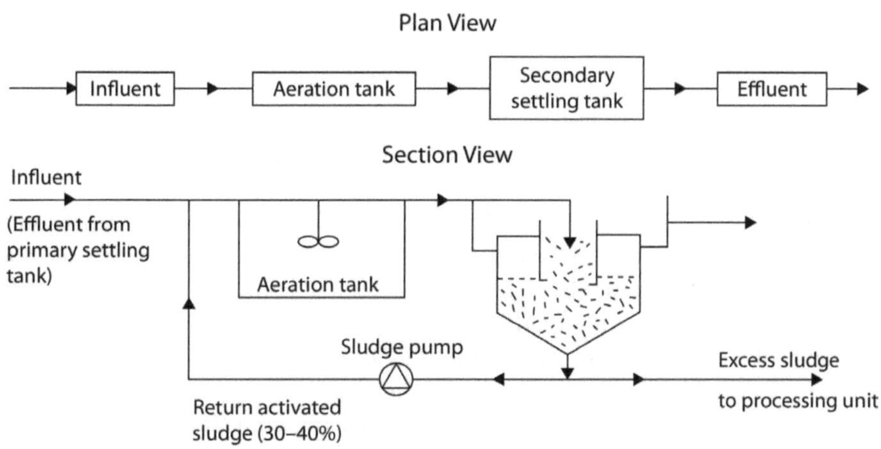

FIGURE 7.5 Flow diagram for activated sludge process.

(Permission granted courtesy of Laxmi Publications Pvt. Ltd.)

- "A **trickling filter:** or biofilter consists of a circular basin or tower filled with support media such as broken stones, slag, plastic rings, modular plastic fills, etc. (see Figure [7.6]). Wastewater is distributed over the media continuously. Microorganisms become attached to the media to form a biological slime layer (biofilm or microbial slime). Organic matter in the wastewater diffuses into the biofilm, where it is stabilized. Oxygen is supplied to the film by the natural flow of wastewater through the media, depending on the relative wastewater temperature and ambient air temperature. The thickness of the biofilm increases as new microorganisms grow. Frequently, portions of the biofilm slough off the media. The sloughed biomass is separated from the liquid in a secondary settling tank, and is discharged to sludge processing. Clear liquid from the secondary settling tank is called secondary effluent, and a portion is often recirculated to the trickling filter to maintain constant hydraulic distribution of the wastewater over the filter" and to improve the level of treatment to achieve more stringent discharge limits [Raj 08].

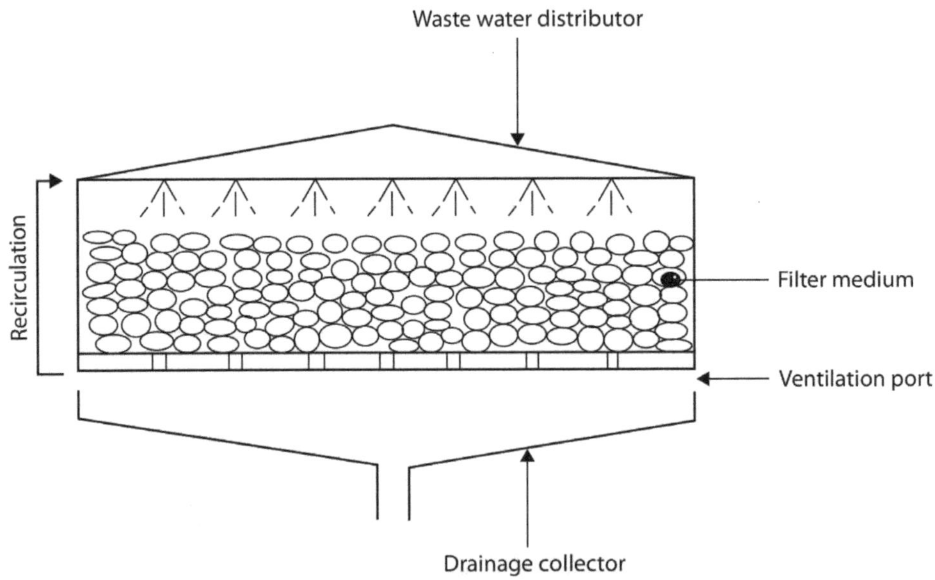

FIGURE 7.6 Schematic representation of conventional trickling filter.

(Permission granted courtesy of Laxmi Publications Pvt. Ltd.)

- **Rotating biological contactors (RBCs):** "(RBCs) are fixed-film reactors similar to trickling filters, in that microorganisms responsible for biodegradation of organic contaminants are attached to support media. In the case of the RBC, the support media are slow rotating discs that are partially submerged in a semicircular tank receiving primary effluent.
A biological contactor consists of a series of closely spaced plastic discs made of poly-vinyl chloride (PVC). The discs are partially submerged in wastewater. By gentle rotation of the discs, the biofilms are alternately exposed to the contaminants in the wastewater and oxygen in the atmosphere. The disc rotation speed affects oxygen mass transfer and maintains the biomass in aerobic conditions. Oxygen is supplied to the attached biofilm from the air

when the biofilm is out of water and from the liquid when submerged. Oxygen is transferred to the wastewater by surface turbulence created by disc rotation. Sloughed pieces of biofilm from the disc surfaces are removed and segregated by providing a secondary settling tank [Figure 7.7].

High-rate biological treatment processes remove not less than 85% of the BOD_5 and suspended solids originally present in raw municipal sewage. Activated sludge process generally produces an effluent of slightly higher quality than a biofilter or RBCs. However, they remove very little phosphorous, nitrogen, and non-biodegradable organics" [Raj 08].

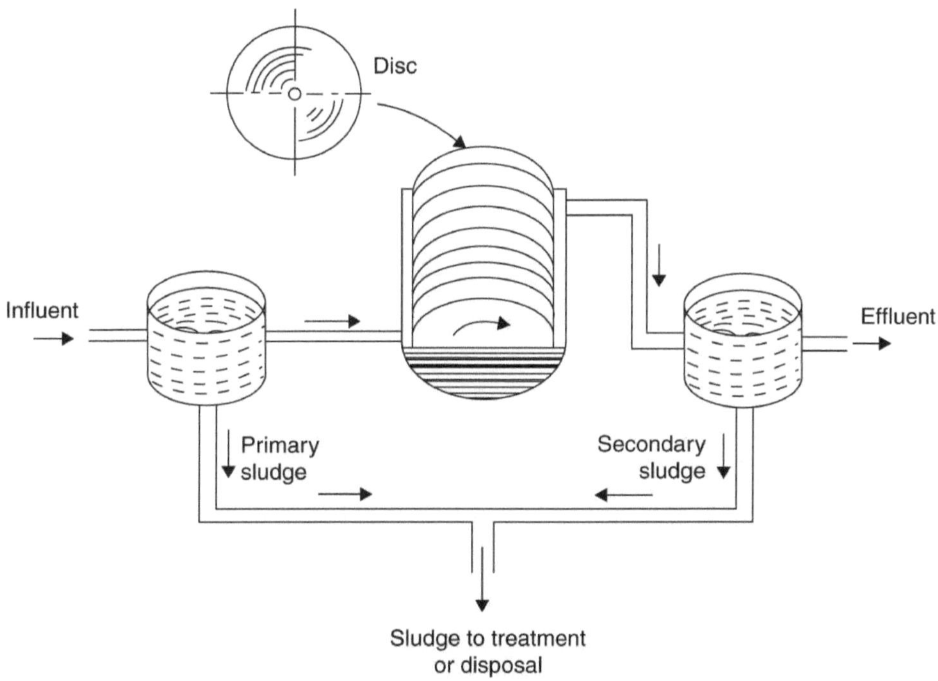

FIGURE 7.7 Schematic representation of rotating biological contactor.

(Permission granted courtesy of Laxmi Publications Pvt. Ltd.)

Low-rate biological treatment systems

"Natural low-rate biological treatment systems are available for the treatment of municipal sewage and tend to be lower in cost and less sophisticated in operation and maintenance, and include: facultative stabilization ponds; aerated lagoons; and batch reactors.

- **Stabilization pond:** shallow ponds, typically 1–2m (3–6ft) deep, where raw sewage or partially treated sewage can be decomposed by symbiotic action of algae and bacteria. These ponds are designed to maintain aerobic conditions throughout, but more often the decomposition taking place near the top layer is aerobic, while the bottom (benthic) layer is anaerobic.

In stabilization ponds, algae utilize carbon dioxide, sulfates, nitrates, phosphates, water, and sunlight to synthesize their own cellular material and give off free oxygen as a waste product. This oxygen, dissolved in pond water, is available to bacteria and other microbes for their metabolic process, which include respiration and degradation of organic material in the pond. The settleable organic matter deposits at the bottom to undergo anaerobic decay [Figure 7.8].

This is the algal-bacterial symbiosis, in which microorganisms use oxygen dissolved in the water and break down organic waste materials to produce end products such as carbon dioxide, water, and plant nutrients. Algae use materials such as nitrates, phosphates, and sulfates as raw material in photosynthesis, to give out oxygen, thus replenishing the depleted oxygen supply and maintaining an aerobic environment" [Raj 08].

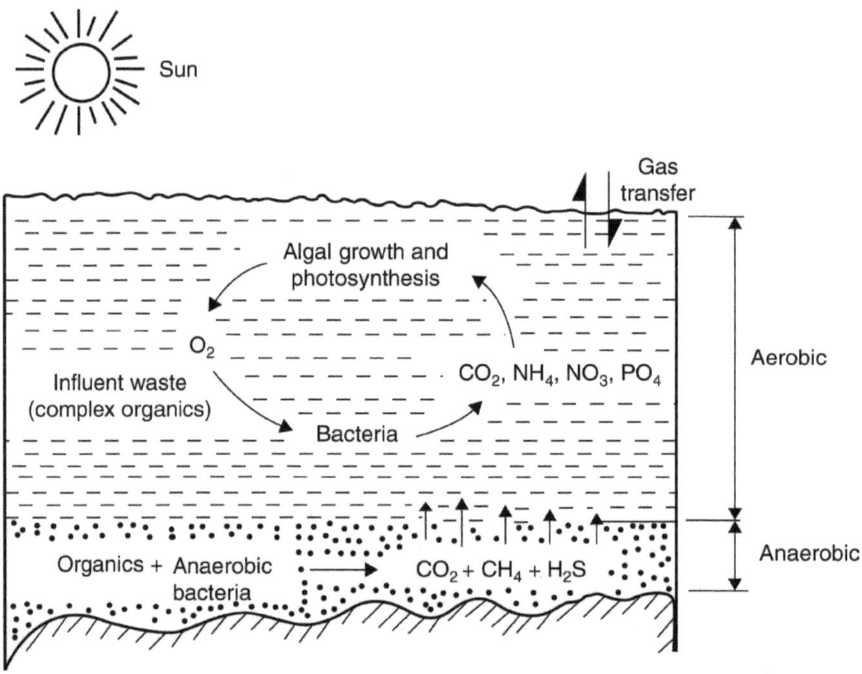

FIGURE 7.8 Stabilization pond.

(Permission granted courtesy of Laxmi Publications Pvt. Ltd.)

- **Aerated lagoon:** "a suspended growth process. The aerated lagoon system consists of a large pond or tank that is equipped with mechanical aerators to maintain an aerobic environment and to prevent settling of the suspended biogas. Initially, the population of microorganisms in an aerated lagoon is much lower than that in an activated sludge system because there is no recirculation of sludge. Therefore, a significantly longer residence time is required to achieve the same effluent quality. The effluent from the aerated lagoon may flow to a settling basin for removal of suspended solids. Alternatively, the mechanical aerators in the system may be shut off for a period of time to facilitate settling prior to discharge of the effluent. The settled solids are generally dewatered prior to disposal" [Raj 08].

Anaerobic Biological Treatment

"Anaerobic treatment converts organic matter in wastewater into a small quantity of sludge and a large quantity of biomass (methane and carbon dioxide), while leaving some pollutants unremoved. In contrast, aerobic processes produce a large quantity of sludge and no biogas, while also leaving some pollution, though less than in the anaerobic treatment process. The main advantages of anaerobic treatment over aerobic treatment are:

- Low operating costs
- Less space requirements; energy recovery (biogas production)
- Low sludge production

Anaerobic stabilization occurs in three stages. In the first stage, complex organics (cellulose, proteins, lipids, etc.) are broken down by extracellular enzymes into soluble organic fatty acids, alcohols, and carbon dioxide. In the second stage, the products of the first stage are converted into various organic acids and alcohols by acid-forming bacteria. In the third stage, one group of methane-forming bacteria converts hydrogen and carbon dioxide to methane, and another group converts acetate to methane and carbon dioxide" [Raj 08]. The anaerobic process is depicted in Figure 7.9.

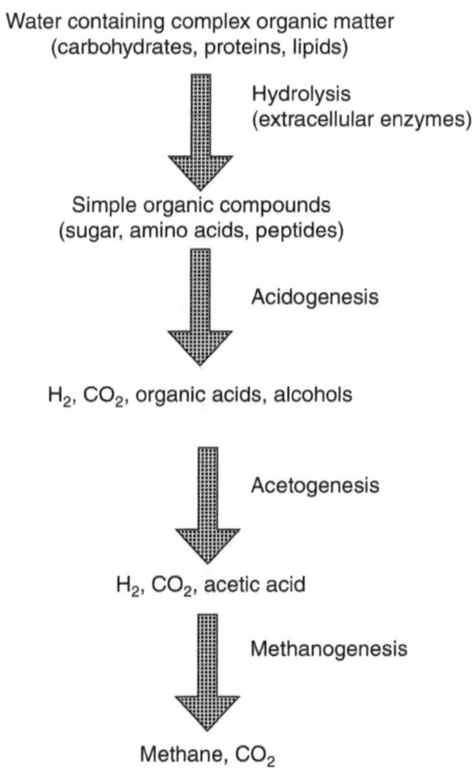

FIGURE 7.9 Microbiology of anaerobic process.

(Permission granted courtesy of Laxmi Publications Pvt. Ltd.)

"The biological degradation of complex organic matter takes place in several consecutive biochemical steps, each performed by different groups of specialized bacteria. In practice, the acetogenic and methanogenic phases are the rate-limiting steps. On the other hand, the generation of methane gas can only happen as fast as methane-forming bacteria receive their substrate. Methane-forming bacteria only use acetic acid (CH_3COOH), hydrogen gas (H_2), and carbon dioxide (CO_2) as substrate.

It is well known that the rate-limiting stage (i.e., the stage that is the slowest) controls the process. In anaerobic processes, methane-forming bacteria reproduce very slowly. These methane-forming bacteria are also very sensitive to environmental factors such as pH, alkalinity, temperature, and toxins. Therefore, controlling these factors is very important to control the speed of the process of digestion.

Anaerobic Digestion of Wastewater Sludges

In primary and secondary treatment processes, a significant fraction of the removed BOD is extracted as sludge. This sludge must be treated further before its safe disposal. One of the most widely employed sludge treatment technologies is anaerobic digestion. Here, a large portion of the organic matter in the sludge is converted to carbon dioxide and methane by microorganisms that act in the absence of oxygen.

The treatment of wastewater sludges consists of two main phases. In the first phase, the objective is to separate the water from the sludge by adopting thickening and dewatering processes. The second phase is known as sludge stabilization. There are three primary objectives of sludge stabilization:

- To reduce the level of pathogens in the residual solids
- To eliminate offensive odors
- To reduce potential for putrefaction

Figure [7.10] shows one of the configurations used for anaerobic sludge digestion. It consists of two identical reactors, with the first one sealed to maintain anaerobic conditions. The second tank may either be sealed for additional gas collection or may be open to the atmosphere. The input for the digester is typically a mixture of solids from the primary settling tank, wasted microbes from the secondary settling tank, and surface scums from both the primary and secondary settling tanks. These sludges are added to the first reactor either continuously or intermittently. This first stage reactor is heated and mixed to accelerate the biological conversion. After a typical residence time of 10-20 days, the mixed digested sludge passes to the second reactor. Here it is retained for further digestion without mixing and heating. Settled sludge is removed from the reactor, either intermittently or continuously. The removed sludge is then dewatered and disposed. The supernatant liquid may be recycled to the beginning of the wastewater treatment plant, if necessary" [Raj 08].

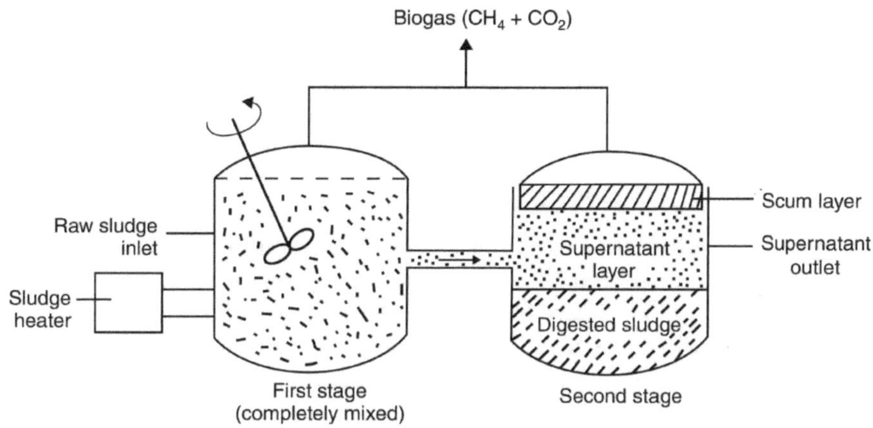

FIGURE 7.10 Schematic representation of anaerobic digester.

(Permission granted courtesy of Laxmi Publications Pvt. Ltd.)

"The significance of the microbiology of anaerobic digestion is manifest in optimum system design and operation. Process design should be directed toward maintaining a large, stable population of methane-forming bacteria. Typically, inadequate system design or operation will result in a relatively high volatile acid accumulation in the digester. The excess acid accumulation may create an imbalance between the population of acid-forming and methane-forming bacteria. In a severe case, excessively high volatile acid production will depress the pH to a level that essentially stops the activity of methanogens entirely. Stable performance can be achieved through careful consideration of the fundamentals of anaerobic treatment. The following environmental and operational parameters are to be considered while designing and operating an anaerobic system:

- **Environmental parameters:** pH, alkalinity, temperature, nutrient content, and toxic compounds
- **Operational parameters:** solids retention time, and substrate characteristics such as; concentration, composition, and biodegradability" [Raj 08]

Facultative processes are a combination of aerobic and anaerobic activity used in the breakdown of organic matter, and are not discussed in detail in this book. "Facultative lagoon systems, like evaporative (ET) systems, are not widely used for onsite wastewater treatment. They are large in size, expensive to build, perform only a portion of the treatment necessary to permit surface discharge or reuse, and produce large concentrations of algae, which negates their use as direct pretreatment before soil infiltration. They have been used in a few states as an alternative system when a subsurface wastewater infiltration system (SWIS) is not feasible, usually to discharge without further treatment to surface waters, which is generally unacceptable under normal circumstances. In some states intermittent discharge lagoons are required. Storage volume is for all cold weather months (four to six months), making the size of these systems too large for most applications" [EPA 11n].

Tertiary/Advanced Wastewater Treatment

Tertiary and/or advanced wastewater treatment is employed when specific wastewater constituents that cannot be removed by secondary treatment must be removed. Individual treatment processes are necessary to remove excess nitrogen, phosphorous, additional suspended solids, refractory organics, heavy metals, and dissolved solids. Because advanced treatment usually follows high-rate secondary treatment, it is sometimes referred to as tertiary treatment.

Dealing with the Biosolids

"In many areas, biosolids are marketed to farmers as fertilizer. Federal regulation (40 CFR Part 503) defines minimum requirements for such land application practices, including contaminant limits, field management practices, treatment requirements, monitoring, recordkeeping, and reporting requirements.

Properly treated and applied biosolids are a good source of organic matter for improving soil structure and help supply nitrogen, phosphorus, and micronutrients that are required by plants. Biosolids have also been used successfully for many years as a soil conditioner and fertilizer, and for restoring and revegetating areas with poor soils due to construction activities, strip mining, or other practices.

Under this biosolids management approach, treated solids in semi-liquid or dewatered form are transported to the soil treatment areas. The slurry or dewatered biosolids, containing nutrients and stabilized organic matter, is spread over the land to give nature a hand in returning grass, trees, and flowers to barren land. Restoration of the countryside also helps control the flow of acid drainage from mines that endangers fish and other aquatic life and contaminates the water with acid, salts, and excessive quantities of metals" [EPA 04, p. 20].

Disinfection

Domestic wastewater contains microorganisms or pathogens that produce human diseases. Processes used to kill or deactivate these harmful organisms are called disinfection. Chlorine is the most widely used disinfectant, but ozone and ultraviolet radiation are also frequently used for wastewater effluent disinfection.

Chlorine

Chlorine kills microorganisms by destroying cellular material. This chemical can be applied to wastewater as a gas, a liquid, or in a solid form similar to swimming pool disinfection chemicals. However, any free (uncombined) chlorine remaining in the water, even at low concentrations, is highly toxic to aquatic life. Therefore, removal of even trace amounts of free chlorine by dechlorination is often needed to protect fish and aquatic life. Due to emergency response and potential safety concerns, chlorine gas is used less frequently now than in the past.

Ozone

Ozone is produced from oxygen exposed to a high voltage current. Ozone is very effective at destroying viruses and bacteria and decomposes back to oxygen rapidly without leaving harmful byproducts. Ozone is not very economical due to high energy costs.

Ultraviolet Radiation

Ultra violet (UV) disinfection occurs when electromagnetic energy in the form of light in the UV spectrum produced by mercury arc lamps penetrates the cell wall of exposed microorganisms. The UV radiation retards the ability of the microorganisms to survive by damaging their genetic material. UV disinfection is a physical treatment process that leaves no chemical traces. Organisms can sometimes repair and reverse the destructive effects of UV when applied at low doses [EPA 04, p. 16].

7.6 U.S. STORM WATER REGULATIONS

While Congress did not deal with storm water in the laws, rules, and regulations before 1990, they passed significant measures to reduce the effect of nonpoint sources and eventually combined sewer overflows (CSOs), a major source of pollution during and after significant storm events. Here is a brief history of the U.S. storm water regulations:

- "In 1990 EPA issued Storm Water Regulation Phase I with implementation to begin in 1992. The Phase I program targeted municipalities over 100,000 people and certain industrial activities.
- In 1992 EPA issued the Storm Water Baseline Industrial General Permit (industrial including construction > 5 acres).
- In 1995 EPA developed the NPDES Storm Water Multi-Sector General Permit for Industrial Activities (MSGP) - did not include construction.
- In 1998 Final Modification of the NPDES Storm Water Multi-Sector General Permit for Industrial Activities and Termination of the EPA Storm Water Baseline Industrial General Permit occurred.
- In 1998 reissuance of NPDES General Permits for Storm Water Discharges from Construction Activities (CGP).
- In 1999 Storm Water Regulations Phase II were developed with implementation set for March 2003.
- October 2000 (NPDES) Storm Water Multi-Sector General Permit for Industrial Activities (MSGP 2000).
- July 1, 2003 NPDES General Permit for Discharges for Large and Small Construction Activities was issued.
- June 30, 2008 NPDES General Permit for Discharges for Large and Small Construction Activities was issued" [EPA 11o].

The storm water pollution control program requires municipalities to develop and implement a program called Municipal Separate Storm Sewer Systems (MS4). This program makes the communities deal with the separation of combined sanitary/storm water sewer systems and their overflow systems (CSOs), prohibits non-storm water discharges, implements erosion and sediment control programs, addresses construction and post-construction runoff, and develops and implements pollution prevention measures in the municipality to minimize discharges.

This is a major initiative nationwide, and the costs are very high, especially because the sanitary sewer systems and storm water systems are combined under most major cities.

Industrial Pretreatment

The National Pretreatment Program, a cooperative effort of Federal, state, POTWs (publicly owned treatment works) and their industrial dischargers, requires industry to control the amount of pollutants discharged into municipal sewer systems. This is critical in the hazardous waste program, where industries often discharge hazardous waste into publicly-owned wastewater treatment plants. It is critical the hazardous constituents be reduced to a level where they will not upset the treatment process at the POTW.

Pretreatment protects the wastewater treatment facilities and its workers from pollutants that may create hazards or interfere with the operation and performance of the POTW, including contamination of sewage sludge, and reduces the likelihood that untreated pollutants are introduced into the receiving waters.

Under the Federal Pretreatment Program, governed by the U.S. Stormwater Regulations, municipal wastewater plants receiving significant industrial discharges must develop local pretreatment programs to control industrial discharges into their sewer system. These programs must be approved by either EPA or a state acting as the Pretreatment Approval Authority. "More than 1500 municipal treatment plants have developed and received approval for a Pretreatment Program" [EPA 11p].

7.7 CASE STUDIES

Illegal wastewater treatment is sometimes difficult to detect, because the facility can give misleading information that is difficult to refute. The following case study illustrates illegal wastewater treatment at a facility that conducted electroplating activities and did not dispose of their wastewater as they claimed.

Case Study

Illegal Wastewater Treatment

At an inspection of a manufacturing facility that electroplated the products it manufactured, the inspector was told by the facility operator that the electroplating wastes were run through a pretreatment program prior to discharge to a publicly-owned treatment system. The inspector completed the inspection, and when he went back to the office, inquired about the discharge. It turned out the discharge was to the facility's residential septic system, which was designed for only human waste. When the owner had the septic system pumped, the contents were hauled to a local publicly owned wastewater treatment plant by a septic hauler who was not licensed to haul hazardous waste. Upon further investigation, the septic system and leach field were highly contaminated with plating chemicals and had to be cleaned up at the owner's expense, together with a fine for an illegal industrial discharge. The three key violations were: treatment of hazardous waste without a permit, illegal underground discharge of industrial waste without a permit, and transporting hazardous waste without a permit.

Note that this facility later became a hazardous waste cleanup site because the contamination was so extensive, the plume of electroplating wastes extended over a five-acre area.

There are many challenges facing wastewater treatment operators. In the following case study, the operator needed to figure out how to treat a toxic contaminant in low concentrations to save the company millions of dollars annually.

Case Study

From the Field: How to Start Treating a Difficult Wastewater

In one large industrial wastewater treatment plant, the industry was using sequencing batch reactors to treat a very toxic chemical in dilute concentrations (2% solution in water). There was a great deal of interest in using wastewater treatment, because the only other form of treatment to dispose of the waste was incineration, and a great deal of energy was required to burn the water along with the chemical.

Because it was almost impossible to develop microbes in the wastewater lab to treat this toxic chemical, the company found a municipal treatment plant that treated very dilute amounts of these chemicals, so they introduced some activated sludge containing the modified microbes, and it worked. The biota in the reactors took a few weeks to adjust to the higher concentration of the chemical, but the plant is successfully treating thousands of gallons of this waste daily, saving the company millions of dollars each year.

In this chapter, you learned about U.S. wastewater regulations, the sources of contamination of wastewaters, how to classify water pollutants, how to characterize wastewater, discovered different kinds of wastewater treatment, read a summary of the U.S. storm water regulations, and read a case study on illegal wastewater treatment. Finally, you read an article from the field on how to start treating a difficult wastewater.

In the next chapter, you will read about solid waste regulations, learn the types and sources of solid wastes, read about municipal solid waste, learn how to estimate quantities of solid waste, how to characterize solid waste, learn about solid waste collection and recycling of solid waste. You will also discover the problems associated with e-waste and read a case study on Love Canal—Lessons Learned.

EXERCISES

1. Name four man-made sources of water pollution.
2. What are three categories of wastewater characteristics?
3. What are the three stages of wastewater treatment?
4. Explain how a trickling filter treats wastewater.
5. What are the similarities of a trickling filter and a rotating biological contactor?
6. What is the main difference between a tricking filter and a rotating biological contactor (RBC)?
7. What is a combined sewer overflow?

REFERENCES

[EPA 04] Primer for Municipal Wastewater Treatment Systems EPA 832-R-04-001 September 2004, online at *http://water.epa.gov/type/watersheds/wastewater/upload/Primer-for-Municipal-Wastewater-Treatment-Systems.pdf*, (accessed May 2011).

[EPA 11n] USEPA Onsite Wastewater Treatment Systems Technology Fact Sheet 7, available online at http://www.epa.gov/nrmrl/pubs/625r00008/html/tfs7.htm, (accessed May 2011).

[EPA 11o] USEPA Office of Wastewater Management NPDES Stormwater Program online at *http://www.dec.state.ak.us/water/wnpspc/stormwater/fedleghistory.htm*, (accessed May 2011).

[EPA 11p] USEPA NPDES Stormwater Program, available online at *http://cfpub.epa.gov/npdes/home.cfm?program_id=3*, (accessed May 2011).

[Raj 08] Raj, S. Amal. 2008. *Introduction to environmental science and technology.* 42–56. Laxmi Publications Pvt, Ltd.

[WIKI 11f] Wikipedia.com. Clean Water Act, online at *http://en.wikipedia.org/wiki/Clean_Water_Act*, (accessed May 2011).

CHAPTER 8

SOLID WASTE MANAGEMENT

In This Chapter

- Solid waste regulations
- Types and sources of solid wastes
- Municipal solid waste
- Estimate quantities of solid waste
- Solid waste characterization
- Solid waste collection
- Recycling of solid waste
- Problems associated with e-waste
- Case study: Love Canal—Lessons Learned

As discussed in Chapter 1, "A Brief History of Hazardous Waste," the USEPA defined solid waste in very complicated regulatory terms, and hazardous waste is a subset of solid waste. Although the USEPA definition of solid waste and therefore hazardous waste includes solids, liquids, and compressed gases, discussions in this chapter will be restricted to solid and hazardous wastes that are in the solid phase only.

8.1 SOLID WASTE REGULATIONS

"Congress enacted the Solid Waste Disposal Act of 1965 to address the growing quantity of solid waste generated in the United States and to ensure its proper management. Subsequent amendments to the Solid Waste Disposal Act, such as RCRA, have substantially increased the federal government's involvement in solid waste management.

Numerous portions of this chapter have been quoted directly with permission from Dr. S. Amal Raj, Introduction to Environmental Science and Technology. Laxmi Publications Pvt. Ltd. 2008. 57–73.

During the 1980s, solid waste management issues rose to new heights of public concern in many areas of the United States because of increasing solid waste generation, shrinking disposal capacity, rising disposal costs, and public opposition to the siting of new disposal facilities.

These solid waste management challenges continue today, as many communities are struggling to develop cost-effective, environmentally protective solutions. The growing amount of waste generated has made it increasingly important for solid waste management officials to develop strategies to manage wastes safely and cost-effectively.

RCRA encourages environmentally sound solid waste management practices that maximize the reuse of recoverable material and foster resource recovery. Under RCRA, EPA regulates hazardous solid wastes and may authorize states to do so. Nonhazardous solid waste is predominately regulated by state and local governments. EPA has, however, promulgated some regulations pertaining to nonhazardous solid waste, largely addressing how disposal facilities should be designed and operated. Aside from regulation of hazardous wastes, EPA's primary role in solid waste management includes setting national goals, providing leadership and technical assistance, and developing guidance and educational materials.

The Agency has played a major role in this program by providing tools and information through policy and guidance to empower local governments, business, industry, federal agencies, and individuals to make better decisions in dealing with solid waste issues. The Agency strives to motivate behavioral change in solid waste management through both regulatory and non-regulatory approaches" [EPA 11q].

As discussed in detail in Chapter 3 of this text, the USEPA promulgated several revisions to the Solid Waste Regulations. Specifically, on December 10, 2014, the Administrator signed the final revisions to the Definition of Solid Waste Rule, also known as the DSW rule. According to the USEPA, "this represents a major environmental justice milestone by directly addressing impacts to communities, disproportionately borne by minority and low-income populations, from the mismanagement of hazardous materials sent to recycling, that is a central part of the final rule."

It should be noted that individual states have their own solid waste rules and regulations, and many of these states are in still the process of promulgating their state's rules and regulations to become current with the December 2010 USEPA revisions.

For the most part, the 2014 USEPA revisions did not have an effect on this textbook, except to better regulate facilities that recycle certain hazardous wastes. Therefore, the discussion of the revisions to the Solid Waste rules in this text is limited to those areas that directly affect hazardous waste management.

NOTE *States must be at least as stringent as the federal solid waste rules and regulations, but states are allowed to be more stringent than the federal standards. On some cases, states are more stringent than the federal rules, so it is always advisable to check with the home state to be aware of any possible differences.*

Side Note

Solid Waste Regulatory Compliance Checklist

The USEPA has provided a complete RCRA subtitle D solid waste compliance checklist. Protocol for Conducting Environmental Compliance Audits of Facilities Regulated under Subtitle D of RCRA, available online at

http://www.epa.gov/compliance/resources/policies/incentives/auditing/apcol-rcrad.pdf

This document includes regulatory requirements only through March 12, 2000, but it is very useful as a guidepost for federal regulatory compliance.

Readers must also check with their home state requirements to be sure they are in complete compliance with solid waste regulations.

8.2 TYPES OF SOLID WASTE

RCRA defines the term *solid waste* as

- **Garbage**: putrescible (easily decomposable) wastes resulting from handling, preparation, cooking, and serving of food. (e.g., milk cartons and coffee grounds).
- **Refuse** (rubbish): all non-putrescible refuse except ash (e.g., metal scrap, wall board, and empty containers).
- **Sludges**: from waste treatment plants, water supply treatment plants, or pollution control facilities (e.g., scrubber slags).
- **Industrial wastes**: manufacturing process wastewaters and non-wastewater sludges and solids.
- **Other discarded materials**: solid, semisolid, liquid, or contained gaseous materials.
- **Certain wastes**: from industrial, commercial, mining, agricultural, and community activities (e.g., boiler slags).

NOTE *__The RCRA definition of solid waste__ is not limited to wastes that are physically solid. Many solid wastes are liquid, while others are semisolid or gaseous. The term solid waste, as defined by RCRA, is very broad, including regular municipal garbage and hazardous wastes. Hazardous waste is a subset of solid waste, and is regulated under RCRA Subtitle C. (Hazardous waste is more fully discussed in Chapters 1 and 2, "Identification of Hazardous Waste.")*

8.3 SOURCES OF SOLID WASTE

The sources and types of solid waste are summarized in Table 8.1.

TABLE 8.1 Sources and types of solid waste

Sources	Activity	Types of Wastes
Residential	Single family swelling, multifamily dwelling, low, medium, and high rise apartments	Garbage, ashes, special wastes
Commercial	Shops, restaurants, markets, offices, hotels, motels, and institutions	Garbage, ashes, construction and demolition waste, and special wastes
Industrial	Fabrication, light and heavy manufacturing, refineries, power plants	Garbage, chemical, and special waste
Open areas	Streets, alleys, vacant lots, playgrounds, beaches, parks, highways, etc	Garbage and special wastes
Treatment plant	Water, sewage and industrial waste treatment plant	Treatment plant wastes

Source: Permission granted courtesy of Laxmi Publications Pvt. Ltd.

Municipal Solid Waste (MSW)

"Municipal solid waste comprises small to moderately sized solid waste items from residential areas, commercial buildings, and institutions. For the most part, this waste is picked up by general collection trucks or typical compactor trucks, on regular routes. In more rural areas, residents may bring their own waste directly to a solid waste transfer facility, recycling center, or landfill.

Negative Effects of MSW

Improperly managed municipal wastes have the following potential negative effects:

- Promotion of microorganisms that cause diseases
- Attraction and support of disease-transmitting vectors like flies and rodents
- Generation of obnoxious odor
- Degradation of the aesthetic quality of the environment
- Occupation of space that could be used for other purposes
- Pollution of the environment" [Raj 08]
- Wastes Resources

8.4 ESTIMATION OF QUANTITY OF MUNICIPAL SOLID WASTE

"The quantity and general composition of the waste materials that is generated is of critical importance in the design and operation of solid waste management systems. The following two methods are recommended to estimate the quantity and composition of solid waste.

1. **Load-count analysis**: In this method, the volume and general composition of each load of waste delivered to a landfill or transfer station are noted during a specified period of time. The total mass and mass distribution by composition is determined using average density data for each waste category (Table [8.2]).
2. **Mass-volume analysis**: This method of analysis is similar to the above method with the added feature that the mass of each load is also recorded. Unless the density of each waste category is determined separately, the mass distribution by composition must be derived using average density value" [Raj 08].

TABLE 8.2 Typical densities for solid waste components and mixtures

Item	Bulk Density (kg/m^3)
Food wastes	290
Paper	85
Cardboard	50
Plastics	65
Textiles	65
Rubber	130
Garden trimmings	105
Wood	240
Glass	195
Tin cans	90
Ferrous metal	320
Non-ferrous metal	160
Dirt, ashes, brick, etc.	480
Municipal solid wastes	
Uncompacted	130
Compacted	300

Source: Permission granted courtesy of Laxmi Publications Pvt. Ltd.

8.5 SOLID WASTE CHARACTERIZATION

"The general purpose of solid waste characterization is to promote sound management of solid waste. Specifically, characterization can determine the following:

- The size, capacity, and design of facilities necessary to manage solid waste
- The potential for recycling wastes
- The potential for composting or incinerating the biodegradable fraction of the waste stream
- The potential and effectiveness of waste reduction programs
- Potential sources of environmental pollution in the waste

Physical Characteristics

Two very important physical characteristics for solid waste analysis are moisture content and density.

Moisture Content

The moisture content is expressed on a wet or dry basis. The wet percentage moisture (P_w) of solid waste is equal to the mass of moisture divided by the total wet mass of the solid. The dry percentage moisture (P_d) of solid waste is equal to the mass of moisture divided by the dry mass of the solid.

Drying is usually accomplished in an oven at 77°C for 24 hours to ensure complete dehydration and yet avoid undue vaporization of volatile materials. The usual method for calculating moisture content is:

$$P_w = (w/S_w) \times 100$$
$$P_d = (w/S_d) \times 100,$$

where

$$w = \text{mass of moisture } (S_w - S_d)$$
$$S_w = \text{initial mass of sample}$$
$$S_d = \text{final mass of sample.}$$

Density

The density of mixed solid waste is influenced by the degree of compaction, moisture content, and component composition. As shown in Table [8.2], individual components of municipal solid waste have different bulk densities. The most important use of the knowledge of the density of solid waste is for determination of its compacted volume. Densities of solid waste may be expressed on either an as-compacted or as-discarded basis. The ratio of the as-compacted density ρ_c to the as-discarded density ρ_d is called the compaction ratio r.

$$r = \rho_c/\rho_d$$

Compacting machines are used to reduce the volume of the solid waste before final disposal. Compaction ratios can vary from 2 to 4. Generally 475 to 594 kg/m^3 can be achieved in landfills with a moderate compaction effect. A poorly compacted landfill can achieve only about 297 kg/m^3 of compacted density. The as-discarded densities of municipal solid waste without compaction may vary from 90 to 180 kg/m^3, with a typical value of 130 kg/m^3" [Raj 08].

FIGURE 8.1 Compacting machine, town of Colonie, NY. (Courtesy of town of Colonie, New York.)

Combustion Characteristics

"Most laboratory analyses are performed on samples of solid waste related to combustion and combustion products. The standard laboratory tests in this category are proximate analysis and heating value.

Proximate Analysis

Proximate analysis is a chemical characterization that determines the amount of some surrogate parameters in place of the true chemical content. Surrogate parameters normally determined in proximate analysis are shown below:

- Moisture content
- Volatile matter
- Fixed carbon
- Ash

The moisture content of solid waste is defined as the material lost during heating the material for one hour at 105°C. Volatile matter is the material driven off as a vapor when waste is subjected to a temperature of 950°C for seven minutes but is prevented from burning. Fixed carbon is the combustible material remaining after the volatile material is driven off. Ash is the residue remaining after combustion.

Heating Value

The heating value of solid waste is measured in kilojoules per kilogram (kJ/kg) and is determined experimentally using a bomb calorimeter. A dry sample is placed in a chamber and burned. The heat released at a constant temperature of 25°C is calculated from heat balance. Because the combustion chamber is maintained at a constant temperature of 25°C, water produced in the oxidation reaction remains in the liquid state. This condition enables the maximum heat release to be determined, and is defined as the higher heating value (HHV).

In the actual combustion process, the temperature of the combustion gas remains above 100°C until the gas is discharged into the air pollution control system. Consequently, the water from the actual combustion process is always in the vapor state. The heating value for actual combustion is termed the lower heating value (LHV). The following equation represents the relationship between HHV and LHV:

$$\text{LHV} = \text{HHV} - [(\Delta \text{H}v) \times 9\text{H}),$$

where,

$$\Delta \text{H}v = \text{Heat of vaporization of water} = 2420 \text{ kJ/kg; and}$$
$$\text{H} = \text{Hydrogen content of combusted material.}$$

The factor of 9 results because 1g mole of hydrogen will produce 9g moles of water (i.e., [16 + 2]/2)" [Raj 08].

8.6 SOLID WASTE COLLECTION

"Collection is the first fundamental function of solid waste management. Solid waste collection refers to the gathering of solid waste from places of generation such as residential, commercial, institutional and industrial areas, as well as public parks and campgrounds. There are three basic methods of collection:

1. Curbside collection
2. Set-out and set-back collection
3. Back yard collection

The quickest and most economical point of collection is from curbs using standard containers. In curbside collection, dwellers should keep the container on curbside at the time of collection. The crew or the truck empty the containers into the collection vehicles and redeposit the containers to the original location. The materials separated for recycling are kept in separate containers, and are picked up by either separate collection trucks or trucks with separate compartments for regular garbage and recyclables.

The set out and set back collection method is sometimes used at larger facilities, and consists of a set out crew carrying the full containers from the storage area to the curb before the collection vehicle arrives, the collector loading the refuse, and the set back crew returning the empty containers.

Backyard collection consists of crews using separate containers or totes to pick up refuse in back yards, bringing the waste to the front for pickup, and the collection vehicle picking up the refuse. This is more convenient for homeowners, but is more expensive" [Raj 08].

Solid Waste Disposal

The USEPA developed a hierarchy of solid waste management to minimize the use of landfills. This hierarchy favors source reduction to reduce both the volume and toxicity of waste and to increase the useful life of manufactured products. The next preferred tier in the hierarchy is recycling, which includes composting of yard and food wastes. Source reduction and recycling are preferred over the third tier of the hierarchy, which consists of combustion and/or landfilling, because they divert waste from the third tier, and because they have positive impacts on the environment and on the economy.

NOTE *EPA did not express a preference for combustion over landfilling as they did in hazardous waste management. They consider combustion and landfills as equal priority. This is likely because municipal wastes break down in a fairly predictable way in landfills, as discussed later in this chapter, while hazardous wastes do not break down in landfills as predictably or as readily as municipal wastes, creating hazardous leachate.*

40 CFR Part 268

Sanitary Landfill

"A sanitary landfill is defined as a land disposal site employing an engineered method of disposing of solid waste on the land in a manner that minimizes environmental hazards by spreading the solid waste to the smallest practical volume, and by applying and compacting cover material at the end of each operational day. Important aspects in the implementation of sanitary landfills include

1. Site selection;
2. Land filling method and operation;
3. Decomposition of the landfilled waste;
4. Emission from the landfills; and
5. Leachate movement and its control.

Site Selection

Site selection is perhaps the most difficult obstacle to overcome in the development of a sanitary landfill. While choosing a location for a landfill, the following parameters are taken into consideration:

- Public opposition
- Proximity of major transportation routes (highways and railways)
- Available land area
- Travel distance
- Soil condition and topography
- Surface water hydrology
- Geological conditions
- Availability of cover material
- Climatologic conditions
- Local environmental conditions
- Ultimate use of site
- State and local ordinances" [Raj 08]

Landfill Methods and Operations

"These are three common structural configurations for landfills:

- Area method
- Ramp method

The area method starts with an excavation either naturally occurring or man-made. After appropriate engineering design and installation of impermeable barriers and leachate collection and detection systems, solid waste is placed on the liner system, spread in layers, and driven over with specially designed compaction equipment. A depth of 2–3 meters is achieved by successive layering of waste. A layer of soil called daily cover is placed over the waste at the end of each operating day. At the end of the operating life of the landfill, final cover is placed, with a properly designed impermeable cap and a gas collection system.

The ramp method is a variation of the area method in which the ground surface is a slope.

Recommended depths of cover for various exposure periods vary, and are given in Table 8.3. A profile of a typical landfill is shown in Figure 8.2" [Raj 08].

TABLE 8.3 Recommended depth of cover

Type of cover	Minimum depth (cm)	Minimum depth (ft)	Exposure (days)
Daily	15	1	<7
Intermediate	30	2	7 – 365
Final	60	3	>365

Source: Permission granted courtesy of Laxmi Publications Pvt. Ltd.

Decomposition of Landfilled Waste

"After the landfill is closed and capped, decomposition of the waste continues, with some physical as well as chemical changes taking place. Physical changes relate primarily to compression of the waste, leading to settling of the landfill. It is estimated that 90% of the settling occurs in the first 5 years, though it may continue for more than 25 years at a slow rate. Landfills with less original compaction obviously tend to settle more.

Bacterial activity also happens in a landfill. Typically, this activity goes through three different stages. In the first stage, there is aerobic decomposition of the waste, with carbon dioxide (CO_2), water (H_2O), and nitrate (NO_3) as the primary products. When the oxygen supply is depleted, the bacterial activity changes. Now the facultative and anaerobic microorganisms digest organic material, producing primarily volatile acids and carbon dioxide (CO_2), resulting in an increase in acidity to a pH of 4–5. Later, methane-producing bacteria (methanogens) begin to predominate, reducing the volatile acids to methane and carbon dioxide and raising the pH to neutral values. The rate of decomposition and therefore the time required for each of these phases depends on a variety of factors, including moisture content, temperature, rainfall, and permeability of the soil cover, among others. The entire process may take decades to complete" [Raj 08].

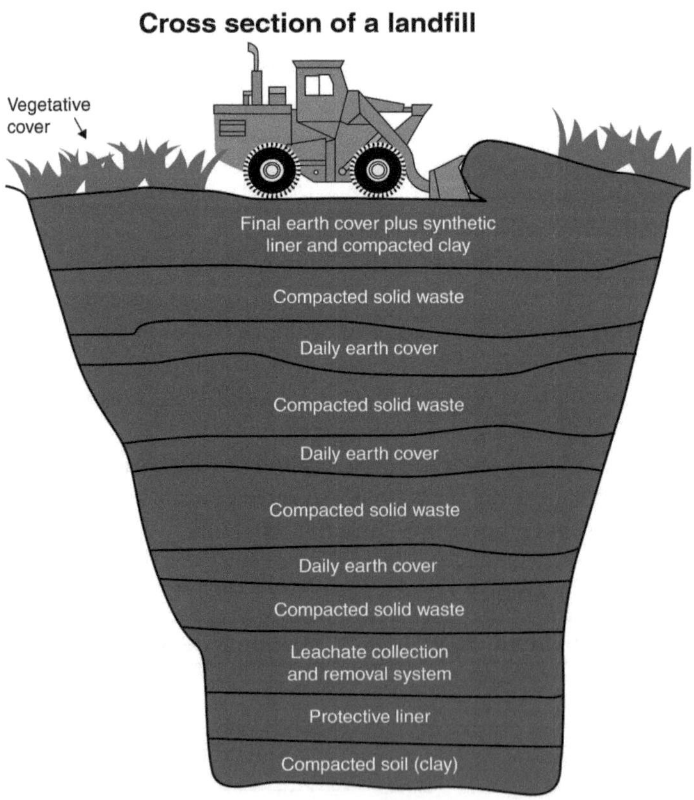

FIGURE 8.2 Sectional view through a sanitary landfill. (Adapted from EPA Quest for Less, educational materials online at *http://www.epa.gov/osw/education/quest/quest.htm.*)

Emissions from Landfills

"Two types of gaseous emissions occur in landfills. Volatiles can permeate out of the landfill through the cap, along the landfill boundaries through the sides, or through the landfill gas collection system. Methane is the most common component of concern, due to its flammability and explosiveness. The other main component of landfill gas is carbon dioxide. In addition to methane and carbon dioxide, minor amounts of nitrogen, hydrogen, hydrogen sulfide, and carbon monoxide have been identified in landfill gas" [Raj 08].

Leachate Movement and Control

"Leachate is liquid seepage from landfills. Leachate from landfills is a documented source of groundwater and surface water pollution. Leachate originates from water percolating through a landfill, picking up soluble constituents. The water that eventually becomes leachate may originate in the waste itself, or from rainwater penetrating the landfill. If untreated leachate leaves the landfill, it may enter either groundwater or surface water and thus act as a vehicle for carrying potentially toxic materials from the landfill to water sources that may be used for human activities. In general, groundwater pollution is more serious than surface water because it is more difficult to detect and clean up. Surface water has some naturally occurring cleaning mechanisms which are lacking in groundwater.

Two approaches have been followed for controlling leachate problems:

1. First, municipal and industrial landfills are required to be designed and constructed with the potential for groundwater pollution in mind. This includes provisions for liners at the bottoms and sides of the landfill to prevent leachate seepage, leachate collection systems to capture leachate before it escapes the landfill, and a leachate leakage detection system, to monitor and detect leaks in the liner and leachate collection system to prevent leaks. Finally, proper construction and operation of the liner, leachate collection system, and operation of the landfill including placement of an impermeable cap over the landfill to prevent precipitation from adding to the leachate.
2. The second approach is the regulation of the types of wastes which are allowed to enter the landfills, limiting the entry of recyclable materials and toxic materials" [Raj 08]. Note that recyclable materials should not be a source of leachate if properly recycled according to the regulations.

Completed (Closed) Sanitary Landfill

"Completed or closed landfills generally require some maintenance because of uneven settling. Maintenance consists primarily of regrading the surface to maintain proper drainage and filling in small depressions to prevent ponding and potential water intrusion. There is also a need to continuously monitor the leachate and leak detection system, along with the gas vents, and gas recovery system, if present. Completed landfills are sometimes used for recreational areas, such as parks, playgrounds or golf driving ranges or golf courses. Of course the landfill gas is an issue, but the collection of gas for energy recovery mitigates this problem, and after a long enough period of time, the biological breakdown of organic materials and thus the generation of landfill gas becomes insignificant" [Raj 08].

Composting

"Composting is an aerobic microbial process which degrades organic matter to produce a relatively stabilized residue, with carbon dioxide as the primary gaseous product. The composting methods may use either manual or mechanical means and are termed as manual and mechanical composting processes. Composting yields a product that contains plant nutrients (nitrogen, phosphorous, and potassium) as well as micronutrients, which can be utilized for plant growth. Prior to composting, salvageable materials are removed, recycled, and reused. The composted material is marketable as a plant nutrient, weed control, and fertilizer, so a major portion of input waste material is reused, resulting in conservation of natural resources. Composting is thus a useful method, especially in predominantly agricultural areas" [Raj 08]. Figure 8.3 illustrates a compost shredder.

FIGURE 8.3 Compost shredder, town of Glenville, NY. (Courtesy of the town of Glenville, New York.)

Process Description

"Most composting operations consist of three basic steps:

- Preparation of the solid wastes
- Decomposition of the solid wastes
- Product preparation and marketing

Receiving, sorting, separation, size reduction, and moisture and nutrient addition are part of the preparation step. To accomplish the decomposition step, the following techniques have been followed:

- Traditional windrow composting
- Aerated static pile composting
- Mechanical composting(in-vessel composting)
- Vermi composting

For each method, the key factors are sufficient moisture and effective distribution of air. Disadvantages include the requirement of relatively large areas on which to place the material to be composted, and unless special precautions are taken, the potential for nuisance due to odor, vermin, insects, rodents, and birds is very high.

Windrow Composting

In windrow composting, prepared solid wastes are placed in long rows of almost triangular cross-section (windrows). [See Figure 8.4] The height is 1–2 m and the base is 3–4 m. The windrows are turned once or twice a week for a composting period of about five weeks. The material is usually cured for an additional period of two to four weeks to ensure stabilization" [Raj 08].

FIGURE 8.4 Composting equipment and windrows town of Glenville, NY. (Courtesy of town of Glenville, New York.)

Aerated Static Pile Composting

The solid waste is piled up in windrows about 1–2 m high, 3–4 m at the base, and 20 m long, laid on floors of ventilated piping systems. To reduce odors, a stabilized compost cover is placed on the piles. The piles are aerated by forcing air through the perforated pipes at regular intervals. Decomposition occurs within four to six weeks. This process has the advantage of better control than the traditional windrow composting.

Mechanical Composting

As an alternative to windrow composting, it is possible to produce humus within five to seven days using mechanical systems. Often the composted material is removed and cured in open windrows for an additional period of about three weeks. Once the solid waste has been converted to humus, it is ready for the final step of preparation and marketing.

This method of composting is carried out in different vessels:

- Horizontal plug flow reactor
- Vertical continuous flow reactor
- Rotating drum

Process Microbiology

Aerobic composting is a dynamic system in which bacteria, acitomycetes, fungi, and other biological forms are actively involved. The relative predominance of one species over others depends on the constantly changing available food supply, temperature, and substrate conditions. In this process, facultative and obligate forms of bacteria, actinomycetes, and fungi are most active. Mesophilic forms are predominant in the initial stages which soon give way to thermophilic bacteria and fungi. Except during the final stages of composting, when the temperature drops, actinomycetes and fungi are confined to 5–15 cm of the outer surface layer. If turning is not carried out frequently, increased growth of actinomycetes and fungi in the outer layer imparts a typical grayish white color.

Attempts are underway to identify the role of different organisms in the breakdown of different materials. Thermophilic bacteria are mainly responsible for breakdown of protein and other readily biodegradable organic matter. Fungi and actinomycetes play an important role in the decomposition of cellulose and lignin" [Raj 08].

"Design Considerations for Composting (Aerobic) Process

- **Particle size**: for optimum results, the particle size should be in the range of 25–75 mm.
- **Seeding**: Composting time can generally be reduced by adding partially decomposed compost (1%–5%) or sewage sludge.
- **Mixing/turning**: To prevent drying, caking and air channeling, materials in the process of being composted should be mixed or turned on a regular schedule.
- **Moisture content**: Should be in the range of 50%–60% during the composting process
- **Temperature**: The optimum temperature for biological stabilization should be between (45°C–55° C) or (113°F–131°F)
- **Carbon-nitrogen ratio**: The initial carbon-nitrogen ratio between 30 and 50 is optimum for the composting process. At lower ratios and higher pH levels, nitrogen is in excess and will be given off as ammonia. At higher ratios, nitrogen will be the limiting nutrient.
- **pH**: The pH should be maintained around 8.5 to minimize the loss of nitrogen in the form of ammonia.
- **Control of pathogens**: At the end of the composting process, the temperature should be maintained between 60°C and 70°C (140°F and 158°F) for 24 hours to destroy pathogenic organisms.

Use of Compost

Compost is beneficial for crop production for the following reasons:

- Compost prepared from municipal solid waste contains on average about 1% each of nitrogen, phosphorous, and potassium (NPK).
- During composting, the plant nutrients are converted to such forms which get released gradually over a longer period of time and do not get leached away easily.
- It also contains micronutrients such as manganese (Mn), copper (Cu), boron (Bo), molybdenum (Mo), etc., which are essential to the growth of plants.
- It is a good soil conditioner and increases the texture of soil, particularly in light of sandy soil.

- It improves the ion exchange and water retaining capacity of the soil.
- The organic matter in soil in tropical climates gets depleted rapidly by microbial activity. Compost adds stabilized organic matter, thus improving the soil.
- It increases the buffering capacity of the soil" [Raj 08].

Anaerobic Digestion

"The process of anaerobic digestion can be described in three phases:

- **Hydrolysis**: The breakdown of high molecular compounds to low molecular compounds as in lipids to fatty acids, polysaccharides to monosaccharides, proteins to amino acids, etc.
- **Acidogenesis**: In this process, low molecular compounds of fatty acids, amino acids, and monosaccharides are converted to lower molecular intermediate compounds such as organic fatty acids.
- **Methanogenesis**: The organic acids are in turn degraded into the final products of methane and carbon dioxide.

In each of these three phases, a different group of bacteria are active and are called the hydrolyzing bacteria, the acidogenic bacteria, and methanogenic bacteria respectively" [Raj 08].

Process Description

"In most processes where methane is to be produced from solid wastes by anaerobic digestion, three steps are involved. The first step is the preparation of organic fraction of the solid wastes for anaerobic digestion and usually includes receiving, sorting, separation, and size reduction. The second step is the addition of moisture and nutrients, blending pH adjustment to 6.7–7.5, heating of the slurry to between 55°C–60°C and anaerobic digestion in a reactor with continuous flow, in which the contents are well-mixed for a period of five to ten days. In most operations, the required moisture content and nutrients are added to the processed solid wastes in the form of sewage sludge. Depending on the chemical characteristics of the sludge, additional nutrients may also have to be added. The third step involves capture, storage, and separation of the gas components evolved during the degradation process. The disposal of the digested sludge is an additional task that must be accomplished" [Raj 08]. A diagram of the process flow for anaerobic digestion is shown in Figure 8.5.

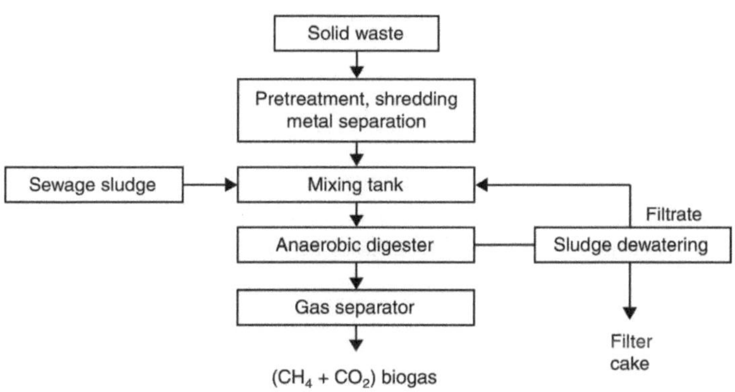

FIGURE 8.5 Process flow diagram for anaerobic digestion. (Permission granted courtesy of Laxmi Publications Pvt. Ltd.)

For anaerobic digestion to successfully occur, it is imperative that the environmental conditions be generated and maintained as follows:

- "The maintenance of an environment that keeps the acidogens and mathanogens in dynamic equilibrium is required to be oxygen-free.
- Should not contain toxic compounds.
- pH should be maintained in neutral range.
- The alkalinity should be in the range 1500–7500 mg/l as $CaCO_3$.
- Have sufficient nutrients such as nitrogen and phosphorous.
- Be temperature steady at either mesophilic or thermophilic conditions.
- Have a constant solids input" [Raj 08].

Pyrolysis

"Pyrolysis is an alternative to incineration for volume reduction and partial disposal of solid waste. The word pyrolysis comes from two Greek words meaning 'fire' and 'breakdown.' Therefore pyrolysis is defined as breakdown by heat. Pyrolysis is an irreversible chemical change brought about by the action of heat in an atmosphere devoid of oxygen. Because most organic substances are thermally unstable, they can, upon heating in an oxygen-free atmosphere, be split through a combination of thermal cracking and condensation reactions into gaseous, liquid, and solid fractions. In contrast to the combustion process, which is highly exothermic, the pyrolytic process is highly endothermic. For this reason, the term destructive distillation is often used as an alternative term for pyrolysis. Destructive distillation of wood to produce methanol and coal gasification are two other common pyrolytic processes.

Normal combustion, as in conventional incineration requires the presence of a sufficient amount of oxygen, which will ensure complete oxidation of organic matter. Using cellulose ($C_6H_{10}O_5$) to represent organic matter, the reaction is:

$$C_6H_{10}O_5 + 6O_2 \rightarrow 6CO_2 = 5H_2O = \text{Heat}.$$

In order to ensure complete combustion and to remove the heat produced during the reaction, excess air is supplied which leads to air pollution problems.

In pyrolysis, the reaction would be:

$$3(C_6H_{10}O_5) \rightarrow 8H_2O + C_6\,H_8O + 2CO = CH_4 + H_2 + 7C$$

$$CO + H_2 \rightarrow HCHO \text{ (formaldehyde)}$$

$$CO + 2H_2 \rightarrow CH_3OH \text{ (methanol)}.$$

When the solid waste is predominantly cellulose, under slow heating at a moderate temperature, the destruction of bond is selective (i.e., the weakest breaking first) and the products are primarily a noncombustible gas and a nonreactive char. On the other hand, when the solid waste is rapidly heated to a high temperature, complete destruction of the molecule is likely to take place. Under intermediate conditions, the system would yield more liquids of complex chemical composition. Normally these two processes are referred to as low-temperature and high-temperature pyrolysis, respectively. Normally, pyrolysis is carried out at temperatures between 500°C–1000°C to produce three component streams.

1. **Gas**: It is a gas mixture containing hydrogen, carbon monoxide, carbon dioxide, methane and some hydrocarbons.
2. **Liquid**: It contains a mixture of tar, pitch, light oil, and low boiling organic chemicals like acetic acid, acetone, methanol, etc.
3. **Char**: Consisting of almost elemental carbon along with inert material that might have entered the process.

The char, liquid, and gas have a large calorific value. This calorific value should be utilized by combustion. Part of this heat obtained by combustion of either char or gas is often used as process heat for the endothermic reaction. It has been observed that even after supplying the heat necessary for pyrolysis, certain amount of excess heat still remains which can be commercially exploited" [Raj 08].

Incineration/Combustion

"Much of solid waste is combustible, and the destruction of this fraction, coupled with energy recovery, is an option in solid waste management. Combustion is a chemical reaction where the elements of fuel are oxidized. In waste-to-energy plants, the fuel is, of course, the solid waste. The major oxidizable elements in the solid waste are carbon and hydrogen. To a lesser extent, sulfur and nitrogen are present. With complete oxidation, carbon is oxidized to carbon dioxide, hydrogen to water, and sulfur to sulfur dioxide. Some fraction of the nitrogen may be oxidized to nitrogen oxides.

$$C + O_2 \rightarrow CO_2$$
$$2H_2 + O_2 \rightarrow 2H_2O$$
$$S + O_2 \rightarrow SO_4.$$

For proper incineration, sufficient air must be supplied to meet the requirements of:

- Primary and secondary combustion
- Turbulence for mixing the air and the solid waste

The combustion reaction is a function of oxygen, temperature, time and turbulence. There must be a sufficient excess of oxygen to drive the reaction to completion in a short period of time. The oxygen is most frequently supplied by forcing air into the combustion chamber with the aid of an air blower. Sufficient time must be provided for the combustion reactions to proceed.

Conventional Incineration

The basic arrangement of the conventional incinerator is shown in Figure. [8.6]. The operation begins with the unloading of solid waste from collection trucks into a storage bin. Storage capacity usually averages about the volume the incinerator can handle in one operating day. An overhead crane is normally used to batch load wastes into the combustion chamber" [Raj 08].

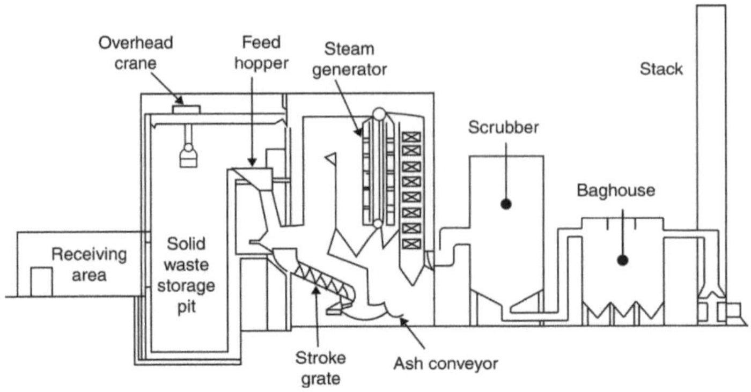

FIGURE. 8.6 Schematic of conventional traveling grate incinerator. (Permission granted courtesy of Laxmi Publications Pvt. Ltd.)

"Although the solid waste may have some heat value, it is normally quite wet and is not autogenous (self-sustaining in combustion) until it is dried. Conventionally, auxiliary fuel is provided for the initial drying stages. In addition to fuel, air may be supplied by means of an air blower.

The combustion chamber consists of bottom grates on which the combustion occurs. As the solid waste enters the combustion chamber and its temperature increases, volatile materials are driven off as gases. Rising temperatures cause the organic component to thermally crack and form gases. When the volatile compounds are driven off, fixed carbon remains. When the temperature reaches the ignition temperature of carbon, it is ignited. Most combustors operate in the range of 980°C–1090°C (1796°F–1994°F), which ensures good combustion and elimination of odors.

The heat liberated during combustion can be recovered by providing a system of pipes in which water is turned to steam to generate electricity.

Some fly ash and other particulate matter may be carried through the chamber. To meet the air pollution control regulations, air pollution devices (i.e. scrubbers, bag houses, etc.) are installed. The cleaned air is discharged through the stack. Bulk volume reduction in incinerators is about 90%. Thus about 10% of the material must still be disposed in a landfill" [Raj 08].

Note that solid waste incineration and the attendant air pollution control measures regulated by the USEPA and authorized states as discussed in Chapter 6.

8.7 RECYCLING OF SOLID WASTE

"Recycling involves using materials that are at the end of their useful lives as the feedstocks for the manufacture of new products. It is differentiated from reuse by the reprocessing and remanufacturing operations" [Raj 08].

NOTE *"Within recycling, a further hierarchy can be defined. Primary recycling is the use of recycled products to make the same or similar products. Secondary recycling is the use of recycled materials to make products with less stringent specifications than the original. This allows for downgrading of the material to suit its possibly diminished properties, and hence is of lower value than primary recycling" [Raj 08].*

All recycling systems must have three major components in order to function. First, they must have a consistent and reliable source of recycled material. Collection of this material is one of the major requirements of this process. Second, method for processing the recovered materials into a form suitable for reuse must be in place. Third, markets must exist for the reprocessed materials in an economically viable manner that a recycling system exists.

The very first step should be separation of solid waste at a domestic level. This varies by municipality, but great progress has been made in residential source separation across the United States. Also, the advent of single stream recycling, where all recyclables can be placed in the same container and separated at the facility, makes recycling easier and more attractive to the customers. For the residents who choose to bring their garbage to a transfer station and those without garbage pickup service, there are recycling centers nearby or at the transfer stations where the recyclables can be dropped off. The success of these recycling centers depends on the prices of the recycled material, and the willingness of the customers to segregate their wastes.

Paper and Cardboard

Postconsumer waste paper is recovered from solid waste in three ways:

1. Source separation of recyclables in households.
2. Salvage industry collection of paper boxes, cardboard boxes, and office papers from industrial and commercial establishments, government and municipal offices, and private collection of paper products
3. Mechanical processing of mixed recyclable waste (single stream recycling)
4. Mechanical processing of mixed municipal solid waste in large recovery plants" [Raj 08].

Figure 8.7 Shows the mechanical processing of solid waste.

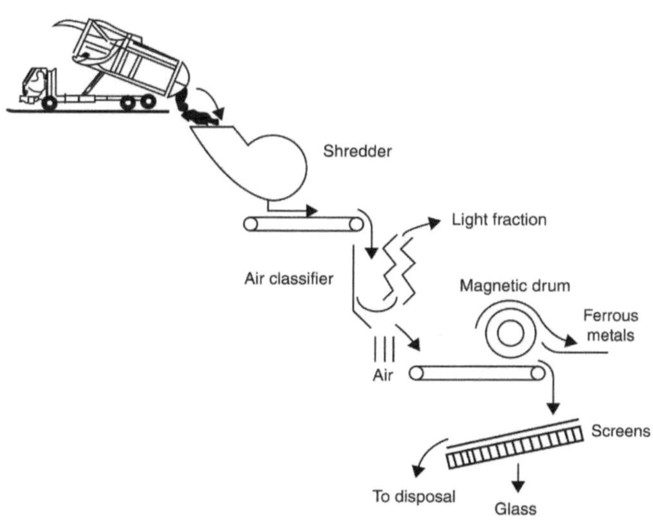

FIGURE. 8.7 Mechanical processing of solid waste. (Permission granted courtesy of Laxmi Publications Pvt. Ltd.)

"Recovered paper can very well be used in the paper industry for manufacturing paper and paperboard products. However, some big paper industries use virgin fiber, while secondary industries often use reclaimed paper for their bulk output. Recycled paper is popular among consumers, and paper manufacturing technologies are making significant advances in the increased use of recycled paper without sacrificing consumer quality.

Tin Cans

Three major potential markets for recycled cans are the steel industries, the de-tinning industries, and the copper precipitation industry.

Aluminum

Aluminum, particularly cans, is a valuable commodity compared to the other common recyclable materials. Recycling of aluminum saves a tremendous amount of energy. It takes 95% less energy to produce an aluminum can from recycling existing cans than from extracting the aluminum from ore" [Raj 08].

NOTE *"Aluminum products that are commonly recycled are siding, gutters, doors, window frames, lawn furniture, and aluminum cans" [Raj 08].*

Glass

Traditionally, glass is well recycled. For recovery from municipal wastes or single stream recycling, glass pieces can be hand-picked or mechanically separated where waste undergoes shredding, air classification, and other types of separation.

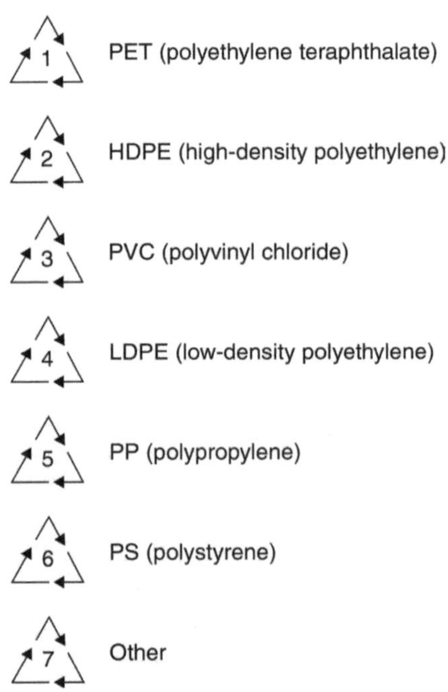

FIGURE 8.8 Basic recycle codes for plastics. (Permission granted courtesy of Laxmi Publications Pvt. Ltd.)

Plastics

The most difficult operation in recycling is the identification and separation of plastics. Because mixed plastic has few economical uses, plastic recycling is only economical if the different types of plastic are separated from each other. The plastic industry has responded by making most consumer products with a code that identifies each type of plastic" [Raj 08]. These codes are pictured in Figure 8.8. Most plastics that can be recycled are common, useful products, some of which are listed in Table 8.4.

TABLE 8.4 Common types of plastics that may be recycled

Code No.	Chemical name	Abbreviated form	Uses
1.	Polyethylene terephthalate	PET	Soft drink bottles
2.	High-density polyethylene	HDPE	Milk cartons
3.	Polyvinyl chloride	PVC	Food packing, wire insulation, pipe
4.	Low-density polyethylene	LDPE	Food wrapping
5.	Polypropylene	PP	Automobile battery casings, bottle caps
6.	Polystyrene	PS	Food packing and eating utensils

Source: Permission granted courtesy of Laxmi Publications Pvt. Ltd.

8.8 ELECTRONIC (E-WASTE)

E-Waste is a popular, informal name for electronic products nearing or at the end of their "useful life." Computers, televisions, VCRs, CD players, DVD players, gaming systems, copiers and fax machines are common electronic products. Unfortunately, discarded electronic items have become part of the fastest growing segments of our nation's waste stream. Certain components of some electronic products contain materials that render them hazardous, depending on their condition and density. Toxic substances contained in electronic products, such as lead, small amounts of precious metals like silver, and non-biodegradable plastics, endanger groundwater under landfills and pose health hazards for neighboring communities. To address solid waste and e-waste issues in the most environmentally sound manner, the three Rs—Reduce, Reuse, and Recycle—should always be applied to all waste streams. We can reduce the amount of e-waste through smart procurement, good maintenance practices, and donating or selling used electronics that are still useful. Only those components that cannot be reused should be recycled. Precious metals recovery is an important part of the recycling, since they have a market value, and are also a serious pollution threat.

Note: For additional information on solid waste from the USEPA, visit the following:

www.epa.gov/epawaste/nonhaz/municipal/wte/index.htm www.epa.gov/wastes/nonhaz/index.htm and www.epa.gov/wastesnonhaz/municipal/msw99.htm

For additional information on composting, visit

www.epa.gov/compost.css.cornell.edu

and

www.calrecycle.ca.gov/swfacilities/

8.9 CASE STUDY

Although this text focuses on hazardous waste, the term hazardous waste was not defined by the USEPA until RCRA was passed in 1976, so all of the landfills created before RCRA were built as solid waste landfills. This includes the Love Canal landfill, which started as a municipal dump for the city of Niagara Falls. The following case study is a discussion of poor solid waste management practices before there were laws to regulate any wastes.

Case Study

Love Canal—Lessons Learned

This is a case study of poor solid waste management. Although there were no rules and regulations in place for waste management when the landfill was closed, the sale of the Love Canal site to developers without the originally attached deed restriction led to many people moving into homes that were too close to a contaminated site.

That said, there are valuable lessons to be learned from the Love Canal Tragedy.

The 990 families that were directly affected by the Love Canal hazardous waste site were not aware of their proximity to a hazardous waste landfill, and had they known about the contamination, would certainly have chosen to live elsewhere. Unfortunately, there have been many other homes built unknowingly near contaminated sites, with the residents potentially suffering adverse health effects. If the contamination at the site is eventually discovered, as in the case of the Love Canal hazardous waste site, they could end up having to leave their homes. Some of these contaminated sites are not necessarily industrial dumps. These contaminated sites may be abandoned gas stations, farms, orchards, or other places where chemicals were placed on the ground by accident or deliberately.

Potential home builders or home buyers should check out the area where they intend to buy or build, and make sure there are no known or potential sources of pollution nearby. If they are considering building a home on a vacant piece of property, they should do a thorough check of the area and make sure the property is clean.

! TIP *Be careful with buying or building homes around old farms and orchards where pesticides, used oil, or other contaminants may be present.*

In this chapter, you read about solid waste regulations and municipal solid waste, learned the types and sources of solid wastes, how to estimate quantities of solid waste, and how to characterize solid waste. Solid waste collection and recycle solid waste were discussed. You discovered the problems associated with e-waste and read a case study on lessons learned by the Love Canal tragedy.

In the next chapter, you will learn about hazardous waste permits, discover exemptions from RCRA permitting, learn how to apply the Land Disposal Restrictions, and read case studies on missing drums, illegal wastewater treatment and illegal discharge, a manufacturing facility making hazardous chemicals and food additives, and a case of source reduction creating waste for which no treatment exists.

EXERCISES

1. Name three solid waste issues that became evident in the 1980s, and continue today.
2. What is EPA's role in nationwide solid waste management?
3. Name the types of solid waste categorized by RCRA.
4. Name the negative effects of solid waste.
5. List and describe the two methods used to estimate municipal waste generation.
6. What is a proximate analysis?
7. What are the key differences between EPA's hazardous waste management hierarchy and their solid waste management hierarchy?
8. Why is electronic waste (E-waste) considered a problem when disposed as other municipal waste?

REFERENCES

[EPA 11q] USEPA Orientation Managing Non-Hazardous Solid Waste, available online at *http://www.epa.gov/osw/inforesources/pubs/orientat/rom2.pdf*, (accessed May 2011).

[Raj 08] Raj, S. Amal. 2008. Introduction to environmental science and technology. 57–73. Laxmi Publications Pvt, Ltd.

CHAPTER 9

HAZARDOUS WASTE PERMITS, TREATMENT, AND TECHNOLOGY

In This Chapter

- Hazardous waste permits
- Exemptions from RCRA permitting
- Apply the Land Disposal Restrictions
- Case studies

A hazardous waste permit for treatment, storage, and/or disposal is rare and very difficult to obtain. These facilities have a negative stigma in the general public, dating all the way back to the discovery of the problems at the Love Canal hazardous waste site (see Chapter 1, "A Brief History of Hazardous Waste"). Despite strong public opposition, industries still generate large quantities of hazardous waste in order to produce many of the products the public demands, so these permits continue to be necessary. A description of hazardous waste permits and their regulatory requirements follows.

9.1 HAZARDOUS WASTE PERMITS

The USEPA requires a facility to obtain a hazardous waste permit for the treatment, storage or disposal of hazardous waste.

FIGURE 9.1 Hazardous waste incinerator. (Adapted from EPA/600/Q-06/002 online at *http://www.epa.gov/ncer/publications/research_results_needs/combustionEmmissionsReport.pdf.*)

A RCRA permit outlines facility design and operation standards, lays out safety standards, and requires the facility to perform many activities such as monitoring and reporting. Permits require facilities to develop emergency plans, provide financial surety, and train employees to handle hazardous waste. Permits also typically require ground-water monitoring, air monitoring, and post-closure funding (financial assurance).

Most of the hazardous waste treatment technologies discussed in Chapter 4, "Hazardous Waste Treatment and Disposal," require the facility to obtain a hazardous waste (RCRA) permit, unless the process is exempted by federal regulations. These exemptions are described in some detail later in this chapter. Obtaining a hazardous waste permit can be a very lengthy and expensive process, and the amount of effort and cost can vary depending on the state where the facility is to be located, the location within the state, and many other factors. Individual states have their own hazardous waste facility siting laws.

If a facility successfully obtains a permit, they are required to provide financial assurance to fund the closure of the facility, to provide corrective action (cleanup) if there is residual pollution, to provide for monitoring after closure of the facility (post-closure costs), and to pay costly permit fees.

Storage of hazardous waste for over 90 days by a Large Quantity Generator (LQG) requires a hazardous waste storage permit. The following case study illustrates a serious (and expensive) violation of this 90-day storage limit.

Case Study

Missing Drums

A hazardous waste inspector went to a facility in response to a report that there were hundreds of 55-gallon containers in the back of the facility. The inspector had called ahead and told the facility there was a complaint and that he would be stopping by in a week.

When the inspector arrived, the facility representative took him for a tour of the facility, only to find a few drums in the accumulation (production) areas, and no wastes in storage at the facility, as reported in the complaint. When the inspector inquired about the alleged large number of drums behind the facility, the representative replied that those wastes had been shipped off site earlier in the week. When the inspector asked for the paperwork (hazardous waste manifests and land disposal restriction certifications), the facility representative said he didn't have any shipping documents because none of the waste had been hazardous.

Coincidentally, the police arm of the state environmental agency had been watching the facility for weeks, and had observed the owner/operator moving a large number of 55-gallon drums of hazardous waste (over three hundred) up to the roof of the building to hide them. The investigators said it was a miracle the roof had not collapsed from the weight of the drums.

After analysis to prove the wastes were indeed hazardous waste, the facility paid a large fine for storing hazardous waste for more than 90 days without a hazardous waste storage permit. The waste had been accumulating for several years, so the cost of properly shipping and managing the waste was added to the fine.

9.2 HAZARDOUS WASTE ACTIVITIES EXEMPT FROM RCRA PERMITTING

There are some treatment activities that are not subject to RCRA permitting. The main exemptions are listed below:

Hazardous Waste Clean-Up Sites

These sites are not normally required to obtain a hazardous waste permit. Hazardous waste permits are usually required for facilities to treat or dispose of hazardous wastes, or to store these wastes on site for more than 90 days. There is one major exception to this general rule, namely hazardous waste cleanup sites. Facilities that are set up to treat wastes generated at hazardous waste cleanup sites are exempt from hazardous waste permitting provided they meet the Applicable or Relevant and Appropriate Requirements (ARARs). These ARARS can be chemical, location, or action specific. When hazardous waste inspectors visit these cleanup sites, it is normally in a cooperative, advisory capacity to provide regulatory guidance only.

Treatment in Wastewater Treatment Units

Since wastewater treatment is normally covered under the Clean Water Act, either under the National Pollution Discharge Elimination System (NPDES), a state run program, or the pre-treatment program, the USEPA decided to regulate hazardous wastewater treatment under the Clean Water Act. The exemption does not apply to the actual waste or the sludges generated during treatment, and the unit(s) must be dedicated to treatment of wastes generated on site.

Treatment in 90-Day and 180-Day Accumulation Areas

As described in Chapter 2, "Identification of Hazardous Waste," and Chapter 3, "Identification of Hazardous Waste," Small Quantity Generators (SQGs) are allowed to exceed the 90-day storage limit under certain conditions without requiring a RCRA storage permit. In addition, generators are allowed to perform some types of hazardous waste treatment in the containers, provided they execute a waste analysis plan to ensure the safety of the treatment.

Recycling

Although recycling a waste is technically treatment, the federal regulations specifically exempt units in which on-site recycling is conducted. The storage of waste prior to recycling is subject to the rules and regulations, as are the resulting wastes and residuals, as well as all air-emission standards.

Elementary Neutralization Units

Elementary neutralization units are units (mainly tanks and containers) used to treat hazardous wastes because they are corrosive. These units are also exempt from generator requirements such as storage time limits (90-day storage) and identifying acceptable waste disposal options for acute wastes. This can represent major cost savings for companies that generate corrosive wastes eligible for neutralization. See Chapters 2 and 3.

Treatment in Totally Enclosed Treatment Facilities

This exemption from hazardous waste permitting is tricky because it applies to units that have two distinct characteristics: they have to be directly connected to an industrial process, and there cannot be any release of any waste or other chemicals during treatment. The wastes in the units are not exempted under CFR §264, 265, and 270.

! TIP *Please note that these rules usually vary from state to state, so it is worthwhile to check with the regulatory in the state hosting the facility before conducting these federally exempt activities.*

The following case study is about a sign manufacturer that committed a combination of serious violations of the hazardous waste and wastewater regulations in two different areas: illegal industrial discharge to a septic system without a wastewater discharge permit; and illegal disposal of an industrial waste to a publicly owned treatment works (POTW) without proper pretreatment and failure to obtain prior approval from the POTW.

Case Study

Illegal Wastewater Treatment and Illegal Discharge to a POTW

In a rural community, a sign shop was inspected for hazardous waste compliance with no advance notice. The facility representative stated that all of the hazardous waste generated at the facility was discharged to a holding tank, which was, in turn, pumped out by a septic tank service. Upon further investigation, the "holding tank" was the facility's septic tank. Some of the industrial waste was running through the tank to the domestic leach field, and the remaining waste was hauled away by an unlicensed waste hauler, who was discharging the industrial hazardous waste into a manhole to a publicly owned wastewater treatment plant. The facility owner was fined for illegally discharging the hazardous waste to his septic system, and the waste hauler was charged with illegal transportation and disposal of hazardous waste. When the facility owner was ordered to dig up his septic system, the entire leach field was contaminated with inks and solvents, and the owner paid thousands of dollars to dispose of the wastes and replace the septic system.

The violations in this case were: transporting of hazardous waste by an unlicensed hauler; illegal discharge of industrial waste into a Publicly Owned Treatment Works (POTW); and illegal underground injection of industrial waste without a permit.

The next case study raises interesting questions, the most pressing being: should a facility that produces industrial products be allowed to also produce food products, food additives, or pharmaceuticals?

Case Study

Manufacturing Facility Making Hazardous Chemicals and Food Additives

Many chemical manufacturers refine and re-refine chemicals to reuse them in their manufacturing processes. However, some chemical manufacturers go to great lengths to recycle their wastes. A hazardous waste inspector visited a small-scale chemical manufacturing facility, and the owner of the facility showed the inspector several specialty chemicals, including additives for automotive antifreeze, food supplements, and other specialty products. He claimed he was recycling all of the hazardous wastes through distillation, concentrating all of the solvents and reusing them, from alcohol to acetone to benzene. When the inspector was finished, the recycling was found acceptable, though it turned out the still bottoms were not being counted as hazardous waste and were being thrown out as municipal waste. This was a minor violation, which was resolved. The concept of making food and other industrial chemicals in the same reaction kettles used for food additives raised issues with food safety, so the inspector alerted the U.S. Food and Drug Administration about his concerns.

9.3 APPLYING THE LAND DISPOSAL RESTRICTION PROGRAM (LDR)

As mentioned earlier, in 1984 Congress was dissatisfied with the amount of hazardous wastes being landfilled without treatment, so they directed the USEPA to develop a hierarchy of hazardous waste management to encourage industries to manage their wastes in a more environmentally friendly manner before using valuable landfill space. Congress also directed the USEPA to develop the Land Disposal Restriction program to make sure hazardous wastes were properly treated to protect human health and the environment prior to the land disposal of any waste [EPA 11r].

The hierarchy chosen at that time was

- Recycling;
- Treatment technologies;
- Solidification/stabilization or immobilization; and
- Land disposal of only treated or solidified/stabilized/immobilized residues.

NOTE *The hazardous waste management hierarchy was updated by Congress in the Pollution Prevention Act of 1990, as described in the beginning of Chapter 3.*

LDR Treatment Standards

The LDR program is a very complex regulatory program, with many complex rulings and nuances. The material concerning LDRs in this text is intended to give the reader a basic understanding of the LDR program. The USEPA provides much more detailed and specific information on their Web site [EPA 01].

As part of the development of the LDRs, Congress directed the USEPA to evaluate each hazardous waste generated, and determine how it should be treated prior to disposal.

The treatment standards for all hazardous wastes were developed by EPA under a very strict time frame, dictated by Congress. As a result, an approach was used whereby technologies were assigned based on the EPA's judgment of the "Best Demonstrated Available Technology" (BDAT) for each waste. The wastes were then run through representative laboratories or existing versions of these technologies, and treatment standards were developed based on the results of these tests. The treatment standard then applies to each waste stream generated. There are three types of treatment standards.

- **Concentration levels:** Most LDR standards set by EPA are specific concentration levels of contaminants. EPA determined these levels by running waste samples through the BDAT process and seeing what concentration level of contaminants remained. That concentration level became the LDR treatment standard. This concentration-based standard allows the facility to choose the type of technology used to treat the waste.
- **Specified technologies:** EPA prescribed particular treatment technologies for some wastes when there was no conventional way to analyze the hazardous constituents, and

there was a technology available that satisfactorily treated the waste. These technologies have to be used unless a different technology is approved by the EPA Administrator. The specified technology standard takes a great deal of latitude away from the facility generating the waste but assures that the proper treatment technology is used.

- **No land disposal:** These wastes were a separate group that EPA determined could be totally recycled, no longer exists, or the wastes can be treated with no resulting residual.

After a treatment standard was determined, the USEPA determined if sufficient treatment capacity was available nationwide. Where sufficient capacity existed, the standard took effect immediately and the waste had to be treated to meet the standards before land disposal. If there was not enough capacity, the effective date was postponed for up to four years, but all of those postponements have expired, so now the standards apply to all wastes subject to the LDRs.

Different Standards for Waters and Non-Wastewaters

NOTE *"EPA created separate and distinct treatment standards between wastewaters and non-wastewaters. A wastewater was defined as wastes containing less than 1% by weight total organic carbon and less than 1% total suspended solids. All other wastes are classified as non-wastewaters" [EPA 11r].*

Exemptions to the LDRs

The LDRs do not apply to Nonregulated Handlers, Conditionally Exempt Small Quantity Generators, and Transporters who only transport hazardous waste without managing it in any other way.

The LDRs do not apply to wastes if they are not land disposed. For instance, wastes from a hazardous waste cleanup are not subject to the LDRs if they are managed on site.

NOTE *The LDRs do apply to Small Quantity Generators, Large Quantity Generators, and Treatment, Storage, and Disposal Facilities (TSDFs).*

Making the LDR Determination

Generators must make the LDR determination for each waste at the point of generation. After making a hazardous waste determination, the generator must assign the appropriate waste code(s), determine if the waste falls into any subcategories, and finally classify the waste by its treatability group.

There is a table in the LDR Treatment Standards §268.40 that lists the treatment standards. Because this table addresses all wastes, it is very lengthy. This table is part of the regulations included on the companion files included with this book. Figure 9.2 shows part of this table, as an example.

§ 268.40

TREATMENT STANDARDS FOR HAZARDOUS WASTES

Waste Code	Waste description and Treatment/Regulatory Subcategory[1]	REGULATED HAZARDOUS CONSTITUENT		WASTEWATERS	NONWASTEWATERS
		Common Name	CAS[2] Number	Concentration mg/1[3]; or Technology code[4]	Concentration in mg/kg[5] unless noted as "mg/1 TCLP"; or Technology code
D001[9]	Ignitable characteristic wastes, except for the § 261.21 (a) (1) High TOC Subcategory.	NA	NA	DEACT and meet § 268.48 standards[8]; or RORGS; or CMBST.	DEACT and meet § 268.48 standards[8]: or RORGS; or CMBST
	High TOC Ignitable Characteristic Liquids Subcategory based on 261.21 (a) (1) – Greater than or equal to 10% total organic carbon. (Note: This subcategory consists of nonwastewaters only.)	NA	NA	NA	RORGS; CMBST; or POLYM
D002[9]	Corrosive characteristic wastes.	NA	NA	DEACT and meet § 268.48 standards[8]	DEACT and meet § 268.48 standards[8]
D002, D004, D005, D006, D007, D008, D009, D010, D011	Radioactive high level wastes generated during the reprocessing of fuel rods. (Note: This subcategory consists of nonwastewaters only.)	Corrosivity (pH)	NA	NA	HLVIT
		Arsenic	7440–38–2	NA	HLVIT
		Barium	7440–39–3	NA	HLVIT
		Cadmium	7440–43–9	NA	HLVIT
		Chromium (Total)	7440–47–3	NA	HLVIT
		Lead	7439–92–1	NA	HLVIT
		Mercury	7439–97–6	NA	HLVIT
		Selenium	7782–49–2	NA	HLVIT
		Silver	7440–22–4	NA	HLVIT
D003[9]	Reactive sulfides subcategory based on 261.23 (a) (5).	NA	NA	DEACT	DEACT
	Explosives subcategory based on 261.23 (a) (6), (7), and (8).	NA	NA	DEACT and meet § 268.48 standards[8]	DEACT and meet § 268.48 standards[8]
	Unexploded ordnance and other explosive devices which have been the subject of an emergency response.	NA	NA	DEACT	DEACT
	Other reactives subcategory based on 261.23 (a) (1).	NA	NA	DEACT and meet § 268.48 standards[8]	DEACT and meet § 268.48 standards[8]
	Water Reactive Subcategory based on 261.23 (a) (2), (3), and (4). (Note: This subcategory consists of nonwastewaters only.)	NA	NA	NA	DEACT and meet § 268.48 standards[8]

FIGURE 9.2 Treatment standards for hazardous wastes. (Adapted from USEPA Rules and regulations §268.40.)

After making the selection from the treatment standard table, the generator must fill out a land disposal restriction form and attach it to the hazardous waste manifest when the waste is shipped out.

9.4 CASE STUDY

The following case study discusses a dilemma caused by moving waste management up the hazardous waste management hierarchy.

Case Study

Source Reduction Created Waste for Which No Treatment Exists

In the frenzy to reduce or eliminate volatile organic chemicals (VOCs) from paints, automotive paint companies have developed a water-based paint with almost no VOCs that is already being used on new cars and trucks. The paint is applied by computers and is literally baked on at the factory.

Discussion:

Although this sounds great, because paint wastes from these new water-based paints are nonhazardous and it seems to fully support the hazardous waste management hierarchy (source reduction, avoiding the generation of hazardous waste is the highest priority), these new paints created problems that no one envisioned during their development:

- If a vehicle is damaged and the body needs repair, it is impossible for auto body shops painting the new or repaired parts to match the paint on the rest of the car. The auto body shop does not have the equipment to bake on water-based paints, and if they did, the heat applied to the new area might affect the original paint on the undamaged part of the vehicle. The only way to get a perfect match in color and texture is to repaint the entire vehicle, which is much more expensive and labor intensive than the past method.
- Although this water-based paint has the benefit of being nonhazardous, there are currently no facilities available to accept this type of waste paint. Fuel blending for energy recovery is not an option because the waste paint has virtually no fuel value, and although incineration is legal, it is expensive compared with the incineration of the older paints containing VOCs.

There is no easy apparent solution to this dilemma, so it will be interesting to watch what happens as the costs of auto repairs on the newer vehicles with the baked-on, water-based paint are expected to climb dramatically in the near future. It may be that this attempt at source reduction is just too expensive in the final analysis, or consumers will have to tolerate imperfect paint matches on their auto-body repairs.

All of the elements required in a LDR form for SQGS, LQGs and Treatment, Storage, and Disposal Facilities are included in the Appendices.

In this chapter, you read about hazardous waste permits and activities that are exempt from permitting, learned the details of the land disposal restrictions (LDRs), and how the LDR treatment standards drive and sometimes even prescribe the treatment required for certain hazardous waste streams. You read interesting case studies on missing drums, illegal wastewater treatment and illegal discharge, a manufacturing facility making hazardous chemicals and food additives, and source reduction creating waste for which no treatment exists.

In the next chapter, you will learn how to make hazardous waste determinations, how to obtain a USEPA identification number, learn hazardous waste accumulation time limits, discover how to fill out hazardous waste manifests, learn the required information on LDR forms. You will also read case studies on a manufacturer with no hazardous waste generation, and a facility with no LDR forms attached to its manifests.

EXERCISES

1. Name some of the basic requirements of permits for all TSDFs.
2. Name the types of treatment that are exempt from hazardous waste permitting.
3. What was the hazardous waste management hierarchy when the LDRs were developed in 1984?
4. How did Congress/EPA modify this hierarchy in 1990?
5. What types of hazardous waste facilities are exempt from LDR requirements?
6. What document must accompany all LDRs in a hazardous waste shipment?

REFERENCES

[EPA 01] Land Disposal Restrictions Summary of Requirements, August 2001, available online at *http://www.epa.gov/osw/hazard/tsd/ldr/ldr-sum.pdf*, (accessed May 2011).

[EPA 11r]. USEPA RCRA Orientation Manual, Land Disposal Restrictions online at *http://www.epa.gov/osw/inforesources/pubs/orientat/rom36.pdf*, (accessed May 2011).

CHAPTER 10

Hazardous Waste Record-Keeping Requirements

In This Chapter

- Make waste determinations
- How to obtain a USEPA identification number
- Hazardous waste accumulation time limits
- How to fill out hazardous waste manifests
- Required information on LDR forms
- Case studies

10.1 WASTE DETERMINATIONS

As discussed at the end of Chapter 2, "Identification of Hazardous Waste," facilities must thoroughly document all waste determinations. This includes facilities that are claiming they are not regulated. One of the first questions a hazardous waste inspector asks when they inspect a facility is: "Have you made waste determinations on all of the waste generated at the facility?" Companies need to ensure that they have the determination(s) stored in an easily accessible, separate file so the inspector can review them. It is critical to document the waste determination(s) thoroughly, including conversations or correspondence from federal, state or local regulators.

Hazardous Wastes Exempted from Regulation for Generators

There are a few wastes that are exempted from the definition of hazardous waste in CFR Part 261.7:

- Those collected for waste identification and treatability studies
- Residues of hazardous waste in empty containers

- Wastes generated in nonwaste treatment manufacturing units until the waste exits the unit where it was generated, unless the unit is a surface impoundment (very rare) or unless the waste remains in the unit more than 90 days after the unit ceases operation [EPA Aug.10]

NOTE *Although not exempted as a hazardous waste, if any of the hazardous wastes generated on site are universal wastes, as described in Chapter 2, and they are managed as universal waste, they do not count in the hazardous waste generation totals for determining the generator categories or for assessing fees. This can be very important, because each level or category of waste generation carries higher regulatory requirements and fees.*

Examples of Hazardous Wastes Generated by Businesses, Industries, and Institutions

Colleges

- Laboratory wastes, such as spent solvents, unused reagents, reaction products, testing samples, and contaminated materials
- Maintenance wastes, such as parts cleaning solvents, lead acid batteries, other types of batteries, waste fuels, paint chips, spent antifreeze, parts cleaning solvent, and shop towels
- Custodial wastes including pressurized aerosol spray cans, fluorescent bulbs, stains, lacquer, thinners, used paint, and discarded pesticides
- Office wastes, such as used electronics, mercury containing devices (thermostats and switches), personal computers, computer monitors, and other electronic wastes

Marinas and Boat Repair Facilities

- Shop wastes, such as parts cleaning solvents, lead acid batteries, waste fuels, spent antifreeze, shop towels, floor dry and other absorbents, refrigerants, and used oil/used oil filters
- Yard wastes, such as used or discarded paint, spray paint cans, paint strippers and paint chips, hull-cleaning chemicals, lacquers, thinners, used paints, used cleaning towels, and discarded pesticides (tri-butyl tin)
- Office wastes, such as used electronics, mercury containing devices (thermostats and switches), personal computers, computer monitors, and other electronic wastes

Equipment Repair Shops

- Acids and bases, toxic wastes, ignitable wastes, plating wastes, and paint wastes

Dry Cleaning And Laundry Facilities

- Still bottoms (residues) from solvent distillation, spent filter cartridges, cooked powder residue, spent solvents, and unused solvents (perchloroethylene)
- Office wastes, such as used electronics, mercury containing devices (thermostats and switches), personal computers, computer monitors, and other electronic wastes

Automobile Engine And Body Repair Shops

- Ignitable wastes, solvent wastes from parts cleaning, acids and bases, paint wastes, spray paint booth filters, shop towels, refrigerants, and air bag propellants
- Office wastes, such as used electronics, mercury containing devices (thermostats and switches), personal computers, computer monitors, and other electronic wastes

Print Shops

- From the printing process, wastes can include acids and bases from etching, silk screening, and cleaning, heavy metal wastes from inks, solvents from cleaning of presses and rollers, toxic wastes from inks and paints, waste inks, unused chemicals
- Photo processing wastes include acid from etching, solvents from cleaning, ignitable wastes, and silver from the fixers and developers
- Office wastes, such as used electronics, mercury containing devices (thermostats and switches), personal computers, computer monitors, and other electronic wastes

Furniture Manufacturing And Refinishing Operations

- Ignitable wastes from strippers and lacquers, toxic wastes from oil based paints and strippers, solvent wastes from cleaning, paint wastes, and possibly electroplating wastes
- Office wastes, such as used electronics, mercury containing devices (thermostats and switches), personal computers, computer monitors, and other electronic wastes

Leather Manufacturing Companies

- Acids and bases from tanning chemicals, ignitable wastes, toxic wastes, solvent wastes, and unused chemicals
- Office wastes, such as used electronics, mercury containing devices (thermostats and switches), personal computers, computer monitors, and other electronic wastes

Vehicle Maintenance Shop Waste

- Acids and bases, solvents from parts cleaning, ignitable wastes such as used fuel and waste oils, toxic wastes, paint wastes, batteries, used oil, unused cleaning chemicals, refrigerants, and air bag propellants
- Office wastes, such as used electronics, mercury containing devices (thermostats and switches), personal computers, computer monitors, and other electronic wastes

Construction And Demolition Contractors

- Ignitable wastes, toxic wastes, solvent wastes, lead contaminated paint chips, paint wastes, used oil, acids, and bases
- Wastes left behind by previous occupants

Hospitals, Nursing Homes, And Assisted Living Facilities

- Medical wastes, such as disinfectants, sample preservatives, and mixed wastes from chemotherapy
- Laboratory wastes, such as silver from developing x-rays and other films, spent solvents, unused reagents, reaction products, testing samples, and contaminated materials
- Maintenance wastes, such as parts cleaning solvents, lead acid batteries, other types of batteries, waste fuels, paint chips, spent antifreeze, parts cleaning solvent, and shop towels
- Custodial wastes including pressurized aerosol spray cans, fluorescent bulbs, stains, lacquer, thinners, used paint, and discarded pesticides
- Office wastes, such as used electronics, mercury containing devices (thermostats and switches), personal computers, computer monitors, and other electronic wastes

These wastes are not all inclusive for each type of business and are not always hazardous wastes, depending on the results of analytical testing.

The following case study illustrates a unique situation wherein a manufacturer fabricated and painted waste containers without generating any hazardous waste.

Case Study

Example of a Manufacturer with No Hazardous Waste Generation

A hazardous waste inspector visited a new factory that manufactured and painted waste containers (Figure 10.1). When the inspector inquired about the types and quantities of hazardous waste generated at the facility, the owner/operator responded there was no hazardous waste generated on site.

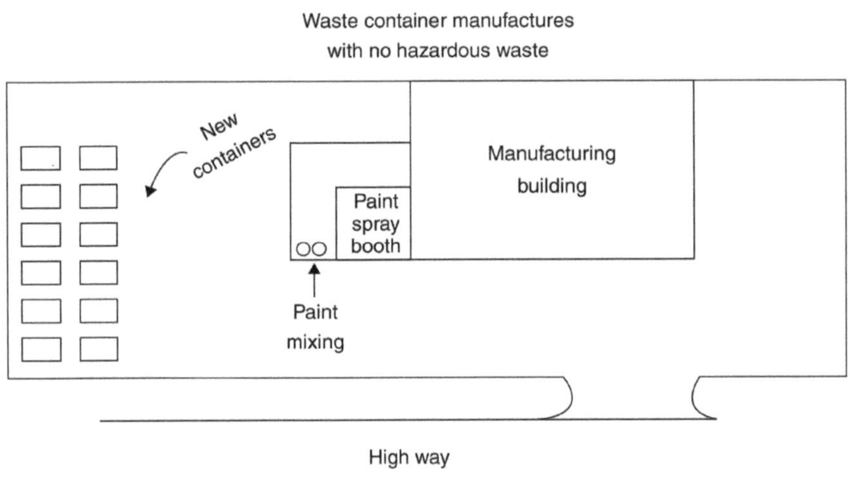

FIGURE 10.1 Waste container manufacturing plant. (Drawing by author.)

During the walk-through portion of the inspection, they inspected the manufacturing areas, the preparation areas, and the paint spray booths. The owner/operator explained that when the paint container was empty, they would add paint thinner to the small amount of paint in the almost empty container and spray it into the new container. If there was any leftover paint thinner mixed with paint at the end of a shift, the workers would paint the mixture into the new containers. As long as they were applying the paint and thinner as paint, they were not generating or disposing of waste.

The used fluorescent lamps were changed out and managed by a building management company, so the universal waste was not counted as waste generated on site. The spray booth filters were new, and did not have to be characterized until they were changed out.

Discussion: Applying the excess paint and leftover thinner to the containers was an effective and legal way to dispose of the excess solvent, and was particularly effective because all of the containers were painted the exact same color. The only future hazardous waste generation would be the paint filters from the spray paint booth. Since it was a new facility, the filters had not yet been replaced. The facility was required to test these filters using the toxicity characteristic leaching procedure (TCLP) before disposing of them.

10.2 USEPA IDENTIFICATION NUMBER

If a facility is an SQG or larger, generating or managing hazardous wastes at a rate greater than 100 kilograms (220 pounds) per month, it must obtain a federal hazardous waste generator identification number (USEPA ID #). This number is unique to the site where the hazardous waste is generated and is issued by the USEPA regional office governing the area where the wastes are generated. The contact page is on the EPA website [EPA 11s].

These USEPA ID numbers are site specific. It is imperative to have an USEPA ID Number for every physical location where hazardous waste is generated at a rate of more than 100 kg per month. Facilities generating less than 100 kg per month are not required to have an USEPA ID number.

10.3 HAZARDOUS WASTE ACCUMULATION TIMES

Generators of hazardous waste need to label their hazardous waste containers correctly, marking the accumulation date and the fill dates. This documentation is important to prove that the waste has been kept in the accumulation areas and < 90-day storage areas within the regulatory time limits. Storage of full containers of wastes in accumulation areas more than three days and in < 90-day storage areas for longer than 90 days are both potential violations. Storing wastes longer than 90 days without a hazardous waste storage permit is very serious, so the containers need to be clearly labeled to prove compliance.

10.4 HAZARDOUS WASTE MANIFESTS

All shipments of hazardous wastes, except universal wastes and shipments by Very Small Quantity Generators (VSQGs), must be accompanied by a completed hazardous waste manifest form. An example is included in Figure 10.2.

Please print or type. (Form disigned for use on elite (12-pitch) typewriter.) Form Approved. OMB No. 2050-0039

UNIFORM HAZARDOUS WASTE MANIFEST	1. Generator ID Number	2. Page 1 of	3. Emergency Response Phone
			4. Manifest Tracking Number

GENERATOR

5. Generator's Name and Mailing Address — Generator's Site Address (if different than mailing address)

Generator's Phone:

6. Transporter 1 Company Name — U.S. EPA ID Number

7. Transporter 2 Company Name — U.S. EPA ID Number

8. Designated Facility Name and Site Address — U.S. EPA ID Number

Facility's Phone:

9a. HM	9b. U.S. DOT Description (including Proper Shipping Name, Hazard Class, ID Number, and Packing Group (if any))	10. Containers No.	10. Containers Type	11. Total Quantity	12. Unit Wt./Vol.	13. Waste Codes
	1.					
	2.					
	3.					
	4.					

14. Special Handling Instructions and Addtional Information

15. **GENERATOR'S/OFFEROR'S CERTIFICATION:** I hereby declare that the contents of this consignment are fully and accurately described above by the proper shipping name, and are classified, packaged, marked and labeled/placarded, and are in all respects in proper condition for transport according to applicable international and national governmental regulations. If export shipment and I am the Primary Exporter, I certify that the contents of this consignment conform to the terms of the attached EPA Acknowledgment of Consent.
I certify that the waste minimization statement identified in 40 CFR 262.27(a) (if I am a large quantity generator) or (b) (if I am a small quantity generator) is true.

Generator's/Offeror's Printed/Typed Name — Signature — Month Day Year

INT'L

16. International Shipments ☐ Import to U.S. ☐ Export from U.S. Port of entry/exit:
Transporter signature (for exports only): Date leaving U.S.:

TRANSPORTER

17. Transporter Acknowledgment of Receipt of Materials
Transporter 1 Printed/Type Name — Signature — Month Day Year
Transporter 2 Printed/Type Name — Signature — Month Day Year

DESIGNATED FACILITY

18. Discrepancy
18a. Discrepancy Indication Space ☐ Quantity ☐ Type ☐ Residue ☐ Partial Rejection ☐ Full Rejection
Manifest Reference Number:
18b. Alternate Facility (or Generator) — U.S. EPAID Number
Facility's Phone:
18c. Signature of Alternate Facility (or Generator) — Month Day Year

19. Hazardous Waste Report Management Method Codes (i.e., codes for hazardous waste treatment, disposal, and recycling systems)

1.	2.	3.	4.

20. Designated Facility Owner or Operator: Certification of receipt of hazardous materials covered by the manifest except as noted in Item 18a
Printed/Typed Name — Signature — Month Day Year

EPA Form 8700-22 (Rev. 3-05) Previous editions are obsolete. **DESIGNATED FACILITY TO DESTINATION STATE (IF REQUIRED)**

FIGURE 10.2 UESPA Uniform Hazardous Waste Manifest. (Adapted from USEPA Uniform Hazardous Waste Manifest Form, online at *http://www.epa.gov/osw/hazard/transportation/manifest/pdf/newform.pdf.*)

10.5 LAND DISPOSAL RESTRICTION (LDR) FORM

All hazardous waste shipments must be accompanied by a completed LDR form. These forms were not created by the USEPA; all waste management companies developed their own versions. The key elements of any LDR form:

- Name and Address of Generator and its USEPA ID Number
- Name and Address of Receiving Facility and its USEPA Number
- Waste description, hazardous waste codes, LDR subcategories at point of generation
- Water or non-wastewater, and underlying hazardous constituents
- Waste disposition, exclusions, waste disposition, technology used, and date shipped (if applicable)
- Was waste discharged under Clean Water Act?
- Was waste characteristic hazardous waste?
- Was the waste "debris" that subsequently became excluded?
- Was the waste "soil" that subsequently became excluded?
- Is the waste residue from treating K061, K062, and/or F006 in high temperature metals recovery?
- Has waste been treated to remove characteristics and to meet underlying hazardous waste constituents standards?
- Was waste treated to remove characteristics but not to meet underlying hazardous waste constituents standards?
- Is the waste debris that has been treated to meet alternative treatment standards?
- Is the waste high-temperature metals recovery residue from treatment K061, K062, and/or F006 wastes?

Dilution of wastes cannot be used as a substitute for treatment. Some exemptions exist for laboratory packs and other wastes.

Case Study

No LDR Forms Attached to Manifests

In the late 1980s, an inspector visited a manufacturing facility that was a Large Quantity Generator (LQG) of hazardous waste. They had done an accurate waste determination and had good record-keeping, with one exception—the facility had no Land Disposal Restriction (LDR) forms attached to their manifests. When the inspector inquired where they were, they responded that the waste hauler had informed them that LDR forms were not necessary. When the inspector explained the requirement and the reasons for the LDRs, they apologized and created the LDR forms immediately. The waste hauler did not violate the generator rules, but did violate the rules by transporting the waste without a LDR form.

The preceding case study illustrates how it can be very unwise for a hazardous waste generator to leave record keeping up to the waste transporter, or to take the advice of the transporter or another company.

! TIP

Facility representatives should always make sure they have accurate, documented waste determinations, accurate and properly prepared hazardous waste manifests and LDR forms for all hazardous waste shipments; they should never rely on anyone other than their own consultation with the USEPA or their home state regulators for waste determinations!

Just as for the requirement that generators make their own waste determinations, they are also responsible for making sure their waste transporters are licensed and all records are correct, including manifests, LDR forms, annual or biennial reports, etc.

In this final chapter you learned how to make waste determinations, how and where to obtain a USEPA identification number, the importance of hazardous waste accumulation time limits, how to fill out hazardous waste manifests and LDR forms. You also read some practical case studies on a manufacturer with no hazardous waste generation and what happened to a facility with no LDR forms attached to manifests.

Using the information contained in this book, you should have a fundamental basic knowledge of the hazardous waste management program in the United States, the basic technical aspects of solid waste management, air pollution control, and wastewater management. You should have a more detailed knowledge of the history of the hazardous waste program, hazardous waste treatment technologies, the laws, rules, and regulations governing hazardous waste management in the United States, and the policies put forward by Congress and the USEPA.

EXERCISES

1. Name a few wastes that are exempted from the definition of hazardous waste in CFR Part 261.7.
2. How might a company that generates some hazardous wastes not have to count its hazardous waste?
3. Name some hazardous wastes that might be generated by a dry cleaner or laundry facility.
4. What level generator must obtain an EPA ID number?
5. Name the most important step in making a hazardous waste determination.

REFERENCES

[EPA Aug.10] USEPA Hazardous Waste Generator Regulations, Version 4, August 2010, online at *http://www.epa.gov/osw/hazard/downloads/tool.pdf*, (accessed May 2011).

[EPA 11s] USEPA Contact Us Page, online at *http://www.epa.gov/epahome/comments.htm*, (accessed May 2011).

Conclusion

Environmental science is a diverse and growing field of study. Within the field of environmental science, hazardous waste management is a challenging specialty, to say the least. This is particularly true in the United States, where the laws have primarily been developed in reaction to events and situations that threatened human health and the environment. The Love Canal tragedy is broadly accepted as the awakening of the United States to the possible adverse consequences of public exposure to hazardous wastes. Ironically, Love Canal was not the largest hazardous waste site discovered in the United States, nor was it found to contain more hazardous chemicals than many other sites. It was the development of the site for residential use that caused the direct and indirect exposure to the toxins that created the problems.

The United States' hazardous waste laws, rules, and regulations are complicated, and the individual states' versions are often more stringent than the federal regulations. Students, professors, and business professionals alike face a daunting task in trying to understand them.

The history of these environmental laws is relatively new compared to many U.S. laws, with the first Solid Waste Law written in 1965. Although the USEPA and individual states have held numerous public information sessions, published many guidance documents, and answered numerous questions, increased knowledge of this program could certainly be a goal of the people who are subject to its regulations.

This book has explained the basic history of the waste management problems that caused Congress to act, and the laws, rules, and regulations that were developed to solve these problems. Also explained were the basic policy elements developed by the USEPA in response to the directions of Congress.

The technologies used to treat and dispose of hazardous wastes were discussed in broad terms, and this book explained how the USEPA applied treatment and technology standards to all hazardous wastes generated.

The reader is encouraged to use this book as a first step in learning the hazardous waste business, including how to identify hazardous wastes, technologies available to manage hazardous wastes, and the rules that govern hazardous waste management.

Given a basic understanding of the history of the hazardous waste management field in the United States, the reader should be able to better understand the technologies, interpret the regulations, and succeed in this challenging field.

These technologies, laws, and regulations serve to protect human health and the environment throughout the United States and are being used as a template for similar controls in other countries across the world. As we progress in the field of waste regulation, it is hopeful that the federal, state, and local governments will continue to better communicate the requirements of these complex programs so that voluntary compliance can be maximized by those regulated parties who want to comply.

It is also imperative that industry and government continue to strive for better emergency preparedness, so that events such as the radioactive releases at Fukushima, Japan, can be minimized; thus, better protection can be afforded for human health and the environment.

Lastly, this third edition was written to explain the proposed federal rules to regulate PFAS and PFOS waste compounds as hazardous substances, along with other proposed federal regulatory changes.

It is highly advisable to keep abreast of these changes, as they can help to avoid trouble with the regulatory community, save your company money if audited, and most importantly, help to protect the environment.

APPENDIX

GLOSSARY

Abandoned materials—materials for which a facility has no further use, and throws away, abandons, or destroys, or intends to do so

Acute hazardous waste—waste considered to present a substantial hazard whether managed properly or not

Adsorption—placing organic contaminants in contact with materials with high surface areas such as activated carbon or charcoal, to cause the contaminants to adhere or adsorb to the surface

Air pollution—presence of substances in such a concentration that it makes the air harmful or dangerous to breathe or causes damage to plants, animals, and the environment

Air sparging—process of injecting air or oxygen through a contaminated soil zone

Air stripping—physical process of transferring volatile organic compounds (VOCs) from aqueous wastes streams into air molecules

ARARS / Applicable or Relevant and Appropriate Requirements—requires cleanups under CERCLA to meet the substantive technical and safety requirements of RCRA

Atomic Energy Act—1946 U.S. law that covers development, regulation, and disposal of nuclear materials and facilities

Biomedical wastes—wastes resulting from medical diagnosis, treatment, immunization, or research activities

Bioreactor landfill—operates to rapidly transform and degrade organic waste

Bioremediation—process of using microbes to clean up harmful chemicals in contaminated media

Boiler—enclosed device that uses controlled flame combustion to recover and export energy in the form of steam, heated fluid, or heated gases

Capping—placing a cover over contaminated materials to keep them from spreading to other areas because of precipitation, wind, or physical contact by humans or animals

Characteristic hazardous waste (D-list)—solid wastes known or tested to exhibit a characteristic trait, such as ignitability, reactivity, corrosivity, or toxicity

Chemical dehalogenation—soils contaminated with halogens (chlorine, bromine, iodine, and fluorine) are dug up, sifted, and crushed, then placed in a reactor on-site with heat and specific chemicals to reduce the halogen content

Circulating wells—ground water is drawn into a well through one screened section and is pumped through the well to a second screened section where it is reintroduced to the aquifer

Clean Air Act (CAA)—law that defines EPA's responsibilities for protecting and improving the nation's air quality

Clean Water Act (CWA)—1972 federal law that set up the National Pollutant Discharge Elimination System (NPDES), creating permits for point sources of pollution

Coagulation—destabilization of particles in a liquid so that they can agglomerate together to produce larger particles

Code of Federal Regulations / CFR—official annual compilation of all regulations and rules promulgated during the previous year by U.S. government agencies, combined with all previously issued regulations and rules of those agencies that are still in effect

Comprehensive Environmental Response, Compensation, And Liability Act (CERCLA)—1980 federal law to clean up abandoned hazardous waste sites

Concrete bunkers and vaults—repositories made of concrete used to store hazardous wastes over long periods of time

Cooperage—facility that builds, restores, or refurbishes cylindrical containers (drums)

Corrosivity—for hazardous waste, includes aqueous wastes with a pH of 2 or lower, or 12.5 or higher, as well as liquids that corrode steel

D-list hazardous waste—see characteristic hazardous waste

Deep-well injection—process that involves using specially designed wells to inject liquid hazardous waste into deep earth strata containing non-potable water

Disposal—discharge, deposition, injection, dumping, spilling, leaking, or placing any waste into the environment, including land, water, or air

Dump—a place or piece of property where municipal garbage or industrial waste is dumped

Electrokinetics—process that treats contaminants in soil by relying upon application of a low-intensity direct current through the soil between ceramic electrodes

Ex-situ / off-site treatment—treatment away from the contaminated site

F-list hazardous waste—solid wastes that are from non-specific industrial sources

Flocculation—agglomeration of coagulated particles, causing coagulated particles to form larger flocs, which can then settle or float faster

Flushing—pumping harmless liquids into wells to push contaminated liquids into recovery wells

Fracturing—using air or liquid to fracture dense soils or rock to enable treatment of contaminants

Garbage—easily decomposable wastes resulting from handling, preparation, cooking, and serving of food

Generator identification number—unique ID number issued by the U.S. EPA to every physical site where hazardous waste is generated

Groundwater cutoff / containment wall—a trench dug around contaminated media and filled with impermeable materials to prevent migration of contaminants from the site

Hazardous and Solid Waste Amendments (HSWA)—1984 U.S. law to develop regulations encouraging alternative waste management in order to reduce the amount of wastes going into landfills

Hazardous waste—waste that poses a risk to human health or the environment and requires special handling and disposal techniques to make it harmless or less dangerous

Hazardous waste generator—person or site whose processes and actions create hazardous waste

Hazardous waste manifest—form used by hazardous waste generators, transporters, and treatment, storage, and disposal facilities (TSDFs) to track all off-site shipments of hazardous waste

Hazardous waste permit—document required for the legal treatment, storage, or disposal of hazardous waste

Hazardous waste transporters—businesses licensed by individual states to transport hazardous wastes, if those hazardous wastes require a manifest

Hazardous waste Treatment, Storage, and Disposal Facility / TSDF—company or person who stores, treats, or disposes of hazardous waste requiring a permit

Hierarchy of hazardous waste management—philosophy that ranks various waste management methods in order of priority

Household hazardous waste —wastes generated in residences by homeowners that are not subject to hazardous waste regulations

Ignitability—liquid waste with a flash point of less than 140°F, and some solid wastes if they spontaneously combust and/or meet certain ignition/burning testing criteria

Incineration—combustion of waste in the presence of oxygen for waste destruction/treatment purposes

Industrial furnace—unit that is an integral part of a manufacturing process and uses thermal treatment to recover materials or energy

Inherently waste-like—materials U.S. EPA considers too hazardous to be recycled because they are considered highly toxic and require special care in handling

In-situ / on-site treatment—treatment conducted at the contaminated site

Internal Revenue Service—bureau of the U.S. Treasury department responsible for collecting taxes

Inventory control—decision to avoid making products that won't be used as a means to facilitate hazardous waste source reduction

Ion exchange—chemical treatment process used to remove dissolved ionic species from contaminated aqueous streams

K-list hazardous waste—process waste from specific industrial manufacturing processes

Land Disposal Restrictions (LDRs)—rules that require hazardous wastes to be treated before disposal on land in order to destroy or immobilize hazardous constituents that might migrate into soil and ground water

Landfills—repository facilities built aboveground to hold municipal or industrial wastes

Land treatment unit—property that uses naturally occurring soil microbes and sunlight to treat hazardous waste

Large Quantity Generator of Hazardous Waste / LQG—facility that generates ≥ 1,000 kg of hazardous waste per month, or greater than 1 kg acute hazardous waste per month, and stores more than 1,000 kg of hazardous waste and more than 1 kg of acute hazardous waste

Listed hazardous waste—see D-list, F-List, K-list, P-list, and U-list

Love Canal—abandoned hydroelectric energy supply canal in a neighborhood near Niagara Falls, NY, which became a municipal and industrial dump

Material elimination—decision whether a particular material is necessary in a process to facilitate hazardous waste source reduction

Material substitution—substituting less toxic or less wasteful materials in product manufacturing

Military munitions / waste—ordnance abandoned or removed from storage for disposal and burned, detonated, incinerated, or treated prior to disposal

Mixed wastes—hazardous wastes mixed with low-level radioactive waste from industrial or research work, usually including paper, rags, plastic bags, or wastewater treatment residues

Monitored natural attenuation—using natural conditions and naturally occurring bacteria that is already present, to clean up contaminants

Multiphase extraction—a vacuum system to remove combinations of contaminated groundwater, separate-phase petroleum product, and vapors from the subsurface

Municipal Separate Storm Sewer Systems / MS4—separation of combined sanitary/storm water sewer systems and their overflow systems (CSOs) in communities

Nanotechnology—treatment processes conducted at a nanoscale, one billionth of a meter

Neutralization—mixing chemical substances with differing pH to cause the chemical(s) to become more neutral

New York State Department of Health / NYSDOH—NY State's primary health agency, formed in 1901 to protect the health of its citizens

New York State Department of Environmental Conservation / NYSDEC—NY State's primary environmental agency directed toward protecting and enhancing the environment

Nonregulated Handler of hazardous waste / NRH—facility that does not generate regulated hazardous wastes

Nuclear Regulatory Commission—U.S. commission that formulates policies, develops regulations governing nuclear reactor and nuclear material safety, and issues orders to licensees

Off-specification chemicals—products that fail to meet quality requirements necessary to be used or sold for their intended purpose

Oxidation—reaction in which valence of a substance increases from the loss of electrons

P-list hazardous waste—solid wastes that are unused or off-specification chemicals

Permeable Reactive Barriers—permeable underground wall that treats contaminated liquids as they are pushed through the wall by pumping harmless liquids into wells on one side of the wall

Pesticide—substance or mixture of substances intended for preventing, destroying, repelling, or mitigating any pest or regulating plant or leaf growth

PFAS—a group of per-and polyfluoroalkyl substances, known for their water- and oil-resistant characteristics

PFOA—Perfluorooctanoic Acid, one class of the PFAS substances, widely used in production of nonstick pan coatings, water- and stain-resistant textiles and food packaging, along with firefighting foams

PFOS—Perfluorooctanesulfonic acid, one class of the PFAS substances used to make products resistant to stains, grease, soil, and water, and firefighting foam

Phytoremediation—using plants to treat contaminated soils

Picric Acid / Trinitrophenol—an unstable acid used primarily as a chemical reagent and as a booster to detonate other, less sensitive explosives, such as TNT (trinitrotoluene)

Pollution—contamination that adversely alters the natural physical, chemical, or biological properties of the environment

Pollution prevention—reducing or eliminating waste at the source (of generation) by various methods

Pollution Prevention Act—1990 U.S. law requiring U.S. EPA to develop and coordinate a pollution prevention strategy, and to develop source-reduction models

Polychlorinated Biphenyls (PCBs)—class of organic compounds with 1 to 10 chlorine atoms attached to a molecule composed of two benzene rings (biphenyl)

Precipitation—the dropping of heavy metals contained in a solution by gravity, usually caused by a change in pH

Primary air pollutants—directly emitted into the atmosphere and found there in the same form

Process modification—changing process parameters such as reaction temperatures, mixing times and methods, and residence times

Product elimination—evaluation of whether a specific product is necessary in order to facilitate hazardous waste source reduction

Pyrolysis—the thermal decomposition of molecules in the absence of oxygen

Reactivity—unstable under "normal" conditions, and can cause explosions, toxic fumes, gases, or vapors when heated, compressed, or mixed with water

Reclamation—using unused or unsold products and byproducts to produce new, useful products

Recycling—using waste materials to manufacture a useful product

Reduction—chemical reaction whereby the valence of a substance decreases from gaining electrons

Reprocessing—running bad, used, or off-specification products back through a process to produce a new, useful product

Resource, Conservation, and Recovery Act / RCRA—1976 U.S. law focused on hazardous waste, which created a "cradle to grave" tracking system for hazardous waste, requiring hazardous waste generators to pay fees for investigation and cleanup of abandoned hazardous waste sites

Reuse—using a waste product as a substitute for a new material to make new useful products

Rubbish—all nonputrescible refuse except ash

Salt formations and underground caves—geologic repositories used to store untreatable wastes for an extended period

Secondary air pollutants—formed in the atmosphere by the interaction of two or more primary pollutants by processes such as photochemical reaction, hydrolysis, and oxidation

Slurry-phase bioremediation—soils are excavated and mixed in water to form a slurry to keep solids suspended and microorganisms in contact with the soil contaminants

Small Quantity Generator of Hazardous Waste / SQG—facility that generates more than 100 kg and no more than 1,000 kg of nonacute hazardous waste per month, or no more than 1 kg of acute hazardous wastes per month

Soil Vapor Extraction / SVE—applying a vacuum to soil to induce the controlled flow of air and to remove volatile and some semi-volatile organic contaminants from the soil

Soil washing—contaminants adsorbed onto fine soil particles are separated from bulk soil on the basis of particle size, in a water-based system

Solidification / stabilization—adding chemicals or other materials to stabilize contaminated media that is difficult to treat by other means

Solid-phase bioremediation—soils placed in a cell or building and tilled with added water and nutrients, then placed on land farm or composted

Solid waste—any solid, liquid, or contained gaseous material that is discarded by being disposed of, burned or incinerated, or recycled

Solid Waste Disposal Act / SWDA—1965 U.S. law that outlined responsible methods for managing household, municipal, commercial, and industrial wastes

Solvent extraction—method that uses an organic solvent to separate organic and metal contaminants from soil

Source reduction—combination of avoiding waste generation, generating the least amount, lowest concentration and toxicity possible

Superfund Amendments and Reauthorization Act / SARA—1984 U.S. law that amended CERCLA and increased the authority and scope of the Superfund program

Surface impoundments—natural topographic depressions, manmade excavations, or diked areas formed primarily of earthen materials and used for temporary storage or treatment of liquid hazardous waste

Thermal treatment—using steam, hot water, hot air, electrical resistance, radio frequency, or thermal conduction to heat and destroy contaminants in soil

Toxic Substances Control Act / TSCA—1976 U.S. law directed the U.S. EPA to require reporting, record-keeping and testing requirements, and restrictions relating to chemical substances and/or mixtures

Toxicity—containing concentrations of certain substances in excess of regulatory thresholds which are expected to cause injury or illness to human health or the environment

Treatment—process that modifies the physical, chemical, or biological character of a waste

Treatment concentration levels—LDR standards set by EPA with specific concentration levels of contaminants

Treatment specified technologies—prescribed by the U.S. EPA for wastes with no conventional way to analyze the hazardous constituents, but technology was available to treat the waste

Treatment, Storage, and Disposal Facilities—entities with appropriate federal and/or state permits to treat, store, or dispose of hazardous waste

U-list hazardous wastes—solid wastes that are discarded chemical products, manufacturing chemical intermediates, and off-specification commercial chemical products that contain certain ingredients, or any soil or other media contaminated by these chemicals

United States Department of Transportation—federal agency responsible for regulating hauling of wastes on highways across the United States

United States Environmental Protection Agency / U.S. EPA—federal agency created in 1970 to facilitate effective governmental coordination of actions that occur on behalf of the environment

Universal waste—wastes that are generated by most businesses and industries on a regular basis, with relaxed rules and regulations to encourage and reduce the cost of management

Very Small Quantity Generator of Hazardous Waste (VSQG)—facility that generates very small amounts of hazardous waste (≤ 100 KG or 220 LB) of non-acute hazardous waste or ≤1 KG or 2.2 LB of acute hazardous waste per month

Vitrification—uses an electric current to melt contaminated soil at elevated temperatures of 2,900°F to 3,650°F

Waste piles—noncontainerized piles of solid, nonliquid hazardous waste that are used for temporary storage or treatment

Index

Bei Fragen zur Produktsicherheit wenden Sie sich bitte an:
If you have any questions regarding product safety,
please contact:

Walter de Gruyter GmbH
Genthiner Straße 13
10785 Berlin
productsafety@degruyterbrill.com